The Darwinian Delusion

The Darwinian Delusion

The Scientific Myth of Evolutionism

Michael Ebifegha

authorHOUSE®

AuthorHouse™
1663 Liberty Drive
Bloomington, IN 47403
www.authorhouse.com
Phone: 1-800-839-8640

First published by AuthorHouse   09/30/2011

ISBN: 978-1-4634-0385-0 (sc)
ISBN: 978-1-4634-0384-3 (hc)
ISBN: 978-1-4634-0383-6 (ebk)

Library of Congress Control Number: 2011907896

Printed in the United States of America

Any people depicted in stock imagery provided by Thinkstock are models, and such images are being used for illustrative purposes only.

Certain stock imagery © Thinkstock.

This book is printed on acid-free paper.

Scripture quotations marked **NIV** *are taken from THE HOLY BIBLE, NEW INTERNATIONAL VERSION®. Copyright © 1973, 1978, 1984 by International Bible Society. Used by permission of Zondervan Publishing House, Grand Rapids, Michigan 49530. All rights reserved.*

Scripture quotations marked **NLT** *are taken from the Holy Bible, New Living Translation, Copyright © 1996. Used by permission of Tyndale House Publishers, Inc., Wheaton, Illinois 60189. All rights reserved.*

Contents

Books by Michael Ebifegha

The Death of Evolution: God's Creation Patent & Seal

The Darwinian Delusion

Creation or Evolution?: Origin of Species in Light of Science's Limitations and Historical Records

Dedicated to the Almighty God,

WHO,

as

ex-atheist Antony Flew

stipulates in his last will and testament,

IS

a self-existent, immutable, immaterial,
omnipotent, and omniscient Being.

Ultimately it comes down to the alternative: What came first? Creative Reason, the Creator Spirit who makes all things and gives them growth, or Unreason, which, lacking any meaning, strangely enough brings forth a mathematically ordered cosmos, as well as man and his reason. The latter, however, would then be nothing more than a chance result of evolution and thus, in the end, equally meaningless.

—Pope Benedict XVI, *Creation and Evolution* (San Francisco: Ignatius Press, 2007).

PREFACE

Life, like love, is immaterial but dwells in and operates through a material body. If the origin of life is unsolvable as a scientific problem, so also is the origin of species. For biological evolution by natural selection in the form of bacteria-to-bacteria or finches-to-finches, the problem of origin is unimportant in that the progeny are variants of their progenitors. Evolution under these circumstances falls within the scientific domain and is the cornerstone of medical applications for the prevention and treatment of human diseases such as SARS, the development of new agricultural products, and the production of industrial innovations. No dogmatic assumptions are required, and scientists understand the mechanisms involved. This mode of evolution governed by artificial selection existed before the Darwinian paradigm of evolution by natural selection modified it. However, in biological evolution by natural selection of the bacteria-to-human type, the question of origin falls outside the domain of scientific investigation. Here the mechanisms are unknown; doctrinaire assumptions are required; and conclusions are distorted by philosophical preferences. Biological evolution by natural selection is a belief described in this book as evolutionism.

The controversy within the scientific community is not between creation (inventing or producing new things from other things) and evolution (things changing with time), both of which are facts of life and processes in science. According to the National Academy of Sciences, "Biological evolution concerns changes in living things during the history of life on earth."[1] There is, therefore, no actual creation-evolution controversy in science. Instead, the controversy concerns creationism versus evolutionism, which are different beliefs about the origin and diversity of life on earth. The creationist within the scientific community believes in an intelligent, conscious, omniscient Creator, the God who in recorded speech claims credit for having created the world. Creationists accept the limited influence of natural selection as a sorter in the diversity of species. Evolutionists, on the other hand, believe in natural selection as the mindless, unconscious, and sole mechanism behind the emergence of all organisms. Creationism is the religious belief in creation, and evolutionism is the religious belief in evolution.

Creationism and evolutionism are two mutually exclusive beliefs. Both are not scientific because they are not capable of proof. However, one of these worldviews must be a delusion since both cannot be true. My objective is to explain why evolutionism is the delusion.

Advocates of evolutionism present its tenets to the public as scientific facts. According to Peter Grace, "Origins are a matter of history or prehistory; hence no scientific statement, qua science, can be made, since there is no possibility of verification by repetition."[2] Many modern scientists ignore this fact in their presentation of data. For instance, the National Academy of Sciences and Institute of Medicine (NASIM) asserts in *Science, Evolution, and Creationism*:

> [T]he common ancestor of humans and chimpanzees was a species estimated to have lived 6 to 7 million years ago, whereas the common ancestor of humans and the puffer fish was an ancient fish that lived in the Earth's oceans more than 400 million years ago. *Thus, humans are not descended from chimpanzees or from any other ape living today but from a species that no longer exists.* Nor are humans descended from the species of fish that live today but, rather, from the species of fish that gave rise to the early tetrapods.[3] (Emphasis theirs.)

These clearly are not scientific facts based on Grace's criterion of "verification by repetition." NASIM, surprisingly, concurs: "[A]ny belief/event that cannot be tested, modified, or rejected by scientific means does not constitute a part of the processes of science."[4] Accordingly, the premise of fish-to-human or chimpanzee-to-human evolution can be accepted only by faith, since it does not meet the scientific requirement of "testing, retesting, and experimentation."[5] The same rules that disqualify creationism as science invalidate evolutionism as science. Nonetheless, ideological disciples of Charles Darwin continue to propound their preconceptions as reliable science. Such double standards and stonewalling are symptoms of academic delusion.

The Darwinian Delusion continues the anti-pseudoscience dialogue that other authors initiated after the publication of Darwin's *The Origin of Species.* John A. Davison's "The Darwinism Delusion" is the latest contribution. Excerpts from Davison's essay appear below because they provide a concise overview of this dialogue.

It is now 147 years since the publication of Darwin's celebrated "On the Origin of Species," yet not a single species has been observed to be formed through the mechanism he proposed. That mechanism, the natural selection of randomly produced variations, is apparently incompetent to transform contemporary species even into a new member of the same genus. The most intensive artificial selection has also proven to be unable to transcend the species barrier . . . .

How then is it possible for an hypothesis to survive without verification? Both the Phlogiston of chemistry and the Ether of physics collapsed when controlled experiment demonstrated them to be without foundation. Darwinism also has failed to survive the acid (decisive) test of experimental verification. Why then has it persisted?

The reason for this paradox is the subject of this brief essay. It is, as my title indicates, because Darwinism is a delusion. The delusion is that evolution (phylogeny) has proceeded as the result of external causes which can be identified and experimentally manipulated. In my opinion that is impossible because such causes do not now and never did exist. They also do not exist for ontogeny, the development of the individual from the egg. Ontogeny and phylogeny are manifestations of the same reproductive continuum . . . .

Many recent authors have spoken of experimental evolution; there is no such thing. Evolution, a unique, historical course of events that took place in the past, is not repeatable experimentally and cannot be investigated that way.[6]

As ex-Darwinist Gary Parker states, "[T]he real significance of the *Darwinian revolution* was *religious* and *philosophic,* not scientific."[7]

Not all scientists would agree with the facts disproving the Darwinian paradigm because delusion is the denial of experiential knowledge, deductive reasoning, empirical evidence, and scientific limitations. Small wonder, then, on the eve of Darwin's 200th birthday, that articles in scientific journals radically differed in their appraisal of the Darwinian paradigm. For instance, the 24-30 January 2009 issue of *NewScientist* states on its front page, "Darwin Was Wrong[:] Cutting Down the Tree of Life." Darwin's tree of life is the backbone of evolutionism. Graham Lawton, the features editor of *NewScientist*, asserts in his cover story titled "Uprooting Darwin's Tree" that one of the iconic concepts of evolution has turned out to be more fantasy than fact and a figment of our imagination.[8] He contends that the tree metaphor could become biologists' equivalent of Newtonian mechanics.[9] Evolutionary biologist Michael Rose writes, "The tree of life is being politely buried," and stresses that the fundamental view of biology needs to change.[10] In contrast to *NewScientist*'s balanced treatment, the January 2009 special issue of *Scientific American* extolled the wonders of evolution. Headlined "The Evolution of EVOLUTION," the issue portrays evolution as the most powerful idea in science, suggesting how Darwin's insights reshaped the world and our views of life.[11] David M. Kingsley's article "From Atoms to Traits," focusing on Darwin's concept of random variations in organisms, posits that diversity arising from DNA changes can result in complex creatures and cultures.[12] Staff writer Kate Wong's "The Human Pedigree" then presents a family tree and gives no indication of abandoning the tree concept that some evolutionists consider wishful fantasy.[13] Two months later the March 2009 special issue of *Discover* proclaimed the supremacy of Darwinism. Karen Wright's "The Ascent of Darwin" asserts that natural selection, or survival of the fittest, fully explains the diversity and complexity of all living things.[14] Evolutionist Sean Carroll, in his interview with *Discover* editor Pamela Weintraub, contends that evolution has replaced a supernatural explanation of human origins with a naturalistic one.[15] It is evident from these articles that the primary goal of evolutionists is one of brainwashing the public to consider Darwinian evolution as indispensable to modern thinking.

When *NewScientist* asked sixteen eminent biologists to identify the largest gaps remaining in evolutionary theory, their answers, published in the 31 January-6 February 2009 issue,[16] varied, but none mentioned Darwin's tree of life that is being politely buried. This is clear evidence that leading evolutionists will not admit publicly that Darwin was wrong. Nonetheless, the two big questions they would most like to see answered concern the origin of life and the dynamics of natural selection. Niles Eldredge desires to know the ecological context in which selection operates; Steven Pinker is curious about how selection leaves its fingerprint on the genome; and Eörs Szathmáry wonders how evolution by natural selection explains complex thought. These are serious gaps in evolutionary theory of which the public should be informed.

Evolutionists' radically different views of their paradigm reflect both reality and fantasy. What gives probity to the cover story of *NewScientist* is that, while challenging Darwin's tree-of-life concept, it argues that his ideas will prove influential for decades to come. On the other hand, what makes the articles in *Scientific American* and *Discover* symptomatic of intellectual fanaticism is that the demonstrated fallacies of evolutionism militate against their seductive defenses of the Darwinian paradigm.

Prominent among these fallacies is the belief that, based on the Darwinian evolutionary time-scale, dinosaurs became extinct 65 million years ago[17] and that grasslands first appeared about 25 million years ago.[18] Recent empirical evidence reveals, however, that dinosaurs ate grass,[19] a discovery that points to a major fault in the evolutionary time-scale. According to the theory, "If dinosaurs had not disappeared so suddenly at the end of the Cretaceous period (65 million years ago), man would not exist today. For as the dinosaurs became extinct, mammals flourished during the Tertiary period (up to 2 million years ago) and spread all over the world and evolved into many different types."[20] Hence, this fundamental doctrine of evolutionism is wrong. So also is the Darwinian tree-of-life model, which postulates when various species existed and diversified on earth. Because evolutionism is a belief, its tenets override contravening scientific evidence. Such fallacies are indicators of delusion. They also are the reason for the present book.

A survey of 2,060 Britons, timed to coincide with the 200th anniversary of Darwin's birth and the 150th anniversary of *The Origin of Species,* revealed that "Half of British adults do not believe in evolution, with at least 22% preferring the theories of creationism or intelligent design to explain how the world came about."[21] The poll sparked a debate in which some respondents suggested that evolution is badly taught in schools. *The Darwinian Delusion* should provide veridical answers to the creationism-evolutionism controversy.

This is a republished version. Some of the websites I have cited may no longer be available, for it took over three years of extensive research to produce this book. I am grateful to all the authors whose works are quoted. I also am immensely indebted to and thankful for those who opt for scientific truth rather than fiction about our origins as a species.

NOTES

1. "Teaching about Evolution and the Nature of Science" (Washington, D.C.: National Academies Press, 1998). *http://www.nap.edu/openbook.php?isbn=0309063647&page=55*. Retrieved on 12 October 2008.

2. "The Death of Evolution?" *http://www.theotokos.org.uk/pages/creation/pgrace/deathevo.html*. Retrieved on 20 October 2008.

3. *Science, Evolution, and Creationism* (Washington, D.C.: National Academies Press, 2008), p. 24.

4. Ibid, p. 43.

5. Ibid, p. 49.

6. "Darwinism as Delusion (response to Richard Dawkins by John Davison), posted 1, November 2006. *http://www.iscid.org/boards/ubb-get_topic-f-6-t-000675.html* Retrieved on 12 March, 2011. Also quoted in K. Ujvarosy, "Darwinism Delusion Exposed by Professor Emeritus of Biology," *American Chronicle,* 9 January 2007. *http://www.americanchronicle.com/articles/18813*. Retrieved on 16 December 2008.

7. *Creation Facts of Life* (Green Forest, Arkansas: Master Books, 1998), p. 76.

8. "Uprooting Darwin's Tree," *NewScientist*, 24-30 January 2009, p. 34.

9. Ibid. p. 39.

10. Ibid.

11. *Scientific American*, January 2009, front cover.

12. "From Atoms to Traits," *Scientific American*, January 2009, p. 52.

13. "The Human Pedigree," *Scientific American*, January 2009, pp. 60-63.

14. "The Ascent of Darwin," *Discover*, March 2009, p. 36.

15. *Discover*, March 2009, p. 42.

16. "Evolution's Final Frontiers: 16 Challenges for a Modern-Day Darwin," *New Scientist*, 31 January-6 February 2009, pp. 41-43.

17. Robin Rees and Clifford Bishop, *The Way Nature Works* (New York: Macmillan, 1998), p. 302.
18. Ibid, p. 94.
19. Dolores R. Piperno and Hans-Dieter Sues, "Dinosaurs Dined on Grass," *Science*, 310.(5751), 18 November 2005, pp. 1126-1128.
20. Rees and Bishop, *The Way Nature Works*, p. 94.
21. Riazat Butt, "Half of Britons Do Not Believe in Evolution, Survey Finds." *http://www.guardian.co.uk/science/2009/feb/01/evolution-darwin-survey-creationism*. Retrieved on 12 March 2009.

INTRODUCTION

Life, consciousness, mind, and the self can only come from a Source that is living, conscious, and thinking . . . . It is simply inconceivable that any material matrix or field can generate agents who think and act. Matter cannot produce conceptions and perceptions. A force field does not plan or think. So at the level of reason and everyday experience, we become immediately aware that the world of living, conscious, thinking beings has to originate in a living Source, a Mind.[1]

          —Antony Flew with Roy Abraham Varghese

Evolution is the science of change that is observable in nature and empirically and repeatedly demonstrable. It is only under these terms that it becomes scientifically correct to say evolution has occurred. Accordingly, bacteria-to-bacteria evolution has occurred, but the Darwinian bacteria-to-human evolution, has not occurred.

          —Michael Ebifegha

The relatively small changes we do see in species, such as wolves growing a heavier coat in a colder climate, or the beak of the finch adapting to the size of available seeds, tend to *preserve* a species rather than transform it into something brand-new. These small ecological adjustments don't seem to go very far, except in the minds of Darwinists, who view them as exhibiting the entire process that created wolves and birds in the first place.[2]

—George Sim Johnston

The doctrines of evolutionism profoundly influence students irrespective of their backgrounds. I first encountered evolutionist indoctrination during a conversation with my nephew when we met in Baltimore. I was then visiting from Canada. Raised as a Christian, I inquired about his religious beliefs. He confessed that he doubted the concept of a living God because evolution taught otherwise. Our conversation ended with my lesson on the giraffe's neck. I promised him that I would do my homework and present my findings. When approximately ten years later I told him of my book titled *The Death of Evolution*, he requested a copy and promised to follow up with a rebuttal.

Not long afterwards I read an article about biological evolution and its alarming influence on students in Africa. Here Richard Dawkins is right in his argument that evolutionism leads to atheism. To ardent evolutionists anything is possible except God's creation by fiat as recorded in the biblical text of Genesis.

Such evidence of evolutionary theory's captivating power caused me to reflect on my own educational background. I do not remember ever having formally studied the subject. If evolution is about explaining things like the giraffe's long neck, I acquired knowledge of this sort from stories told by illiterates. As children living in an African village without electricity, at night we spread mats and gathered around kerosene lanterns. During these memorable sessions the adults told stories that explained how things came to be—for instance, how and why the tortoise assumed its present form, why some species are dumb while others are clever, and why some crawl on the ground. These stories often made sense by correlating with physical evidence. Therefore, when my nephew presented his views about the long neck of the giraffe, they fitted well with the folktales I had learned in my teens.

When it comes to the subject of human origins, these folktales are criticized as mythical. Small wonder, then, that most scientists denigrate the Genesis account of creation as utterly preposterous; even some theologians claim that it is ancient mythology. Still, the explanations by which evolutionary scientists account for human origins, such as life's springing from non-life, are even more mythical than the ones they seek to replace.

NASIM, for instance, claims that the common ancestor of humans and chimpanzees lived 6-7 million years ago and, the common ancestor of humans and the puffer fish lived 400 million years ago,[3] yet these hypothetical progenitors are unknown. It takes more faith to believe in a series of evolutionary myths than to believe in a single myth of creation! The physician and molecular biologist Michael Denton, comparing the Darwinian theory of evolution with the Genesis account, reaches this conclusion:

> Ultimately the Darwinian theory of evolution is no more or less than the great cosmogenic myth of the twentieth century. Like the Genesis-based cosmology, which it replaced, and like the creation myths of ancient man, it satisfies the same deep psychological need for an all-embracing explanation for the origin of the world which has motivated all the cosmogenic myth makers of the past, from the shamans of primitive peoples to the ideologues of the medieval Church.[4]

Given all the rhetoric that Darwinian evolution is a scientific fact, one expects something more than a myth. What is deeply troubling is that evolutionary scientists adamantly present their side of the argument while too often refusing to compare their worldview with that of creationists. For instance, NASIM seeks to discredit the creationist worldview by claiming, "Common structures and behaviours often demonstrate that species have evolved from common ancestors."[5] This is NASIM's generalization that promotes an exclusively evolutionist worldview. As Mary Midgley remarks in *Evolution as a Religion*, "The theory of evolution is not just an inert piece of theoretical science. It is, and cannot help being, also a powerful folktale about human origins."[6] Another group of scientists that chooses to promote the creationist worldview could look at the same evidence and come up with the explanation, "Common structures and behaviours often demonstrate that species have been created by a common designer." Both worldviews are not falsifiable and, consequently, are equally plausible from a scientific point of view.

We know that organisms are both similar and dissimilar in morphology and behaviour. Thus we can reach another conclusion based on the marked dissimilarities: "Different structures and behaviours demonstrate that species evolved from different ancestors or were created by the same designer or different designers." Evolutionists shy away from focusing on dissimilarities because they illustrate the weakness of their theory, which focuses entirely on homologous similarities. Nobel laureate Sir Ernst Boris Chain argues that we should be more interested in the differences rather than the similarities.[7] In other words, it is simply an issue of which story to tell. In the kind of bacteria-to-bacteria evolution that is demonstrable in the laboratory, the similarity is obvious. In the bacteria-to-human model of evolution, however, scientists emphasize limited similarity in DNA since this approach leads to a preferred philosophical conclusion. Let us consider a couple of examples of how evolutionists manipulate the scientific enterprise.

Consider the belief that life can arise from non-life. Although empirical science discounts this view, evolutionists still posit the spontaneous generation of life since the only other choice is intelligent design, which they dislike on philosophical grounds. Suggesting that, given our present state of knowledge, creation is the only answer, physicist H. S. Lipson, Fellow of the Royal Society, remarks in a letter to *NewScientist*, published on 14 May 1981, that it is distasteful for scientists to reject a theory because it does not fit their preconceived ideas.[8] Natural science should not be a religion.

Another example is the belief that random and mindless processes can transform protozoic slime, over the span of millions of years, into hands, limbs, torso, eyes, brain, face, and skeletal system, enabling in turn the development of abstract intelligence, love, hatred, and spirituality. How ludicrous! Die-hard atheistic scientists such as Richard Dawkins maintain that genes created us, body and mind.[9] In his book titled *God: The Failed Hypothesis*, Victor J. Stenger blankly asserts, "The eye is neither poorly nor well designed. It is simply not designed."[10] Stenger believes that the laws of physics came from "nothing" and laments that his views are not recognized by a consensus of physicists.[11] What would the implications be if the laws of physics were arbitrary?

You could use Stenger's views on the laws of physics to practical advantage. If you find yourself in traffic court for a collision, simply tell the judge that the laws of physics are capricious and, thus, that you do not deserve to be charged. For immediate release hope that the judge turns out to be someone like Dawkins who, on the jacket of Stenger's God: *The Failed Hypothesis*, is quoted as saying, "I learned an enormous amount from this splendid book." Dawkins, however, should recall that Albert Einstein, whom he admires as an illustrious thinker,[12] espoused a philosophical view radically different from Stenger's. According to Einstein, "Everyone who is seriously engaged in the pursuit of science becomes convinced that the laws of nature manifest the existence of a spirit vastly superior to that of men, and one in the face of which we with our modest powers must feel humble."[13] Dawkins therefore will have to make up his mind about whom to believe—Einstein or Stenger.

You also might take another cue from atheist Taner Edis, a physicist colleague of Stenger, who in *The Ghost of the Universe* posits, "We can understand the laws of physics not as expressions of a divine will, but as frameworks for accidents."[14] Therefore, to improve your chances of winning in court, you should go with both books in hand. Edis also asserts, "The complexities of life do not require intelligent design; accidents and blind mechanisms do the trick."[15] In reality, accidents and blind mechanisms produce *chaos*, not meaningful and complex designs.

So, in accordance with Edis's postulate, the next time you have a complex research project that does not seem to work, dump it on the highway to wait for "accidents and blind mechanisms," perhaps in the form of a hurricane, to "do the trick." According to Dawkins, "Given infinite time, or opportunities, anything is possible."[16] The problem with the atheistic solution, however, is that you have to wait millions of years for anything to materialize. To be honest, such pseudoscientific fables concocted by atheists are even more illogical than the folktales of my teenage years. The storytellers should replace the "God of the gaps" with the "delusions of the gaps."

Many in the public eye find these fables unconvincing. For instance, G. K. Chesterton, the famous English novelist and critic, once mused, "It is absurd for the evolutionist to complain that it is unthinkable for an admittedly unthinkable God to make everything out of nothing, and then pretend that it is more thinkable that nothing should turn itself into anything."[17] Former atheist Lee Strobel has written that, the more he analyzed the Darwinian paradigm, the more it appeared to be too far-fetched to be credible. For it to be true, Strobel points out, he would have to believe in several impossibilities: that nothing produces everything, that non-life produces life, that randomness produces fine-tuning, that unconsciousness produces consciousness, and that non-reason produces reason.[18] In the absence of any empirical data to back up these tenets, the Darwinian narrative remains unconvincing.

There are, however, some remarkable differences between conventional folktales and their pseudoscientific variants. Ordinary people circulate folktales within their local communities, but academicians in science classrooms present their fables as fact worldwide. Unlike conventional folktales, the latter are accompanied by interpretations that sometimes involve circular reasoning; they also are supported by circumstantial evidence that cannot be falsified scientifically. In addition, these fables are promulgated with public funding and usually are promoted by the media and defended by courts of law against conflicting views. Ordinary folktales are presented as entertainment, invoking some mythical past to make sense of life, whereas their pseudoscientific equivalents seeks to change people's minds about the basis of life itself.

Evolutionary biology is essentially a collection of fables told by leading modern biologists. Dawkins' books (The Selfish Gene, The Blind Watchmaker, River Out of Eden, Climbing Mount

Improbable, Unweaving the Rainbow, and The Ancestor's Tale)
fit neatly in the this grouping. For example, The Selfish Gene,
according to Denis Noble at Oxford University, is a metaphorical
story as opposed to a properly empirical scientific study.[19]
Harvard University paleontologist Stephen Jay Gould confesses,
"Evolutionary biologists in general are famous for their facility
in devising plausible stories; but they often forget that plausible
stories need not be true."[20] In this way plausible stories can
culminate in academic delusion.

Like folktales, evolutionary narratives justify a preconceived
conclusion. The long neck of the giraffe is a scientific fact, but
how and why it is so is simply the opinion of some scientists. In
this book we shall not credit such pseudoscientific fables. Instead,
we will be concerned with credible narratives that relate to the
origin and diversity of life forms. These involve religious elements
and provide answers about our origins and the meaning of life.
Since the latter are philosophical points, religion and science
are in conflict. Creationists seek answers from both the material
and immaterial world, but evolutionists are committed solely to
materialist answers.

The creationism-evolutionism controversy is not about science;
it is about our choice of a designing instrumentality (God or natural
selection). Depending on the instrumentality that is chosen, the
same scientific data are interpreted differently. For instance, in
the evolutionism camp every aspect of life, whether material or
immaterial, must be explained only in material terms. This means
that anything that cannot be expressed in material terms, such
as love and God, does not exist. In like manner, scientists who
follow the evidence wherever it leads assert that intelligence is
involved, a conclusion consistent with God's revelation of creating
the world through wisdom, understanding, and power. Scientists
who insist on materialist explanations attribute all natural designs
to the blind, random, and mindless processes of evolution. In a
highly secular world most scientists today prefer the construct of
natural selection, and some publicly question God's existence.
No one witnessed events at the beginning of time, however, and
therefore such a paradigm is prone to delusion.

What follows is a brief description of how this book will challenge the delusion of evolutionism. Such proponents discredit creationists in the following ways: discriminatory tactics, ambiguous terminologies, partial arguments, unjustified assumptions, and the exploitation of media support. These means enable advocates of the "new atheism" to proclaim that there is no God. This book will expose the fallacies in their propaganda and argue that the delusion is not about God but about unconditional adherence to the Darwinian theory of evolution.

A. Discriminatory Tactics

In the creationism-evolutionism controversy, the evolutionists prevail over the creationists by subtle propaganda. They stereotype creationism as a religious doctrine while presenting evolutionism as a scientific fact. Creation and evolution are both facts of science. Nevertheless, evolutionists' glossary defines "creationism" and leaves out "creation" in order to imply that the two are synonymous.[21] They also define "evolution" and leave out "evolutionism" to suggest that the two are synonymous as well.[22] However, according to the distinguished atheist philosopher Michael Ruse, "Evolution (evolutionism to be exact) is promulgated as an ideology, a secular religion—a full-fledged alternative to Christianity, with meaning and morality."[23] According to Ruse, evolutionists do not want this fact about evolution disclosed to the public. Here we are indebted to Dawkins for informing us that acceptance of the Darwinian theory of evolution leads to atheism.

Another tactic evolutionists use to discredit creationism is to denigrate some of its tenets. To criticize the creationist worldview, the scientific community focuses on the Genesis account of creation, which is open to various interpretations. The popular approach is to generate conflict between the interpretation of Genesis and scientific data. Two misconceptions are that, according to the Bible, the Earth is 6,000 years old and flat. Some creationists have also misconstrued the Scriptures in their interpretation of Genesis. Evolutionists have used this misinterpretation to lure the public into considering evolution as scientific fact and creation as religious myth.

B. Ambiguous Terminologies

Ambiguous terminologies privilege the evolutionist worldview. The term "species," for instance, is spuriously vague. Evolutionists thus consider the varieties of dogs as artificially created new species in the morphological sense,[24] whereas creationists consider them as belonging to the same species. Professor Ernst Mayr, an advocate for evolutionism, contends, "Even at present there is not yet unanimity on the definition of the term species."[25] Naturalists refer to this predicament as the "species problem."

The meaning and scope of microevolution versus macroevolution is also a major source of friction between creationists and evolutionists. A perspicuous definition of these terms is necessary to specify exactly what aspect of evolution is prone to delusion.

C. Partial Arguments

Scientists focus on the genome in order to reach philosophical conclusions that cannot be falsified as required in the scientific enterprise. The evolutionary worldview's primary emphasis is on DNA similarities; however, the genome and genetic code, unlike the DNA molecule, are immaterial. In addition, knowledge of the genome is far from complete. The immaterial aspect of all living things that science cannot analyze or comprehend is part of the puzzle. Seemingly compelling evidence, therefore, may still be very different from the truth. Science must look beyond DNA before making sweeping conclusions about the true design of the living world.

The Scriptures reveal that human beings were made in the image of God with a special mandate to subdue the earth scientifically and technologically. They thus disclose a spiritual dimension that is foreign to science. This explains why human beings are the only creatures endowed with free will to reject or accept God.

Dawkins has declared that God is a delusion, implying that evolutionism is the correct choice. However, God's claim to have created the universe contradicts the modern belief that it came about accidentally. In presenting this claim, God effectively discredits atheistic assumptions. God's speech to an audience of Jews is the historical evidence that proves His existence. Dawkins neither acknowledges nor disproves this fact in *The God Delusion*. In overlooking this crucial event, he fails to "dispose of the positive arguments for belief [in God] that have been offered throughout history."[26]

A veridical claim, irrespective of the claimant's status, presented before an audience has both moral and legal ramifications. Should we dismiss the historical evidence of God's claim as a religious myth and accept instead the secular myth of evolution by natural selection, or should we argue that the claim presented to the ancient world is not binding on the modern world? It is up to individuals to believe or disbelieve ancient history, the timeless bedrock of modern culture. Dawkins chooses not to believe in God because of the following attributes that he deems repugnant: "The metaphorical or pantheistic God of the physicists is light years away from the interventionist, miracle-wreaking, thought-reading, sin-punishing, prayer-answering God of the Bible, of priests, mullahs and rabbis, and of ordinary language."[27] God's claim to have created the universe provides the basis for an effective rebuttal of natural selection as a creative agency in the evolutionist worldview.

D. Unjustified Assumptions

The fourth approach is to invoke assumptions that cannot be falsified, for they are deliberately chosen to promote the evolutionary worldview. Although atheistic scientists cannot reconstruct events that occurred at the beginning of time, it should not come as a surprise that they nevertheless posit assumptions in support of their worldview.

Either a creating agent or a designing instrumentality must occasion the opportunity for something to evolve into another thing. Evolutionists choose natural selection, a mindless and unconscious process that operates only on already existing entities. Natural selection can produce a limited range of adjustments for survival in a changing environment. The idea of evolution by natural selection is derived from the breeding of animals and plants by artificial selection. And artificial selection controlled by human intelligence has a limit beyond which it becomes counterproductive, marking the limit of natural selection as well. The abrupt and inexplicable gaps in fossil data confirm this fact. It is technically wrong to confuse the roles of a selective process and a creative agency. In order to sustain their worldview, evolutionists are prepared to claim the impossible. Denton accordingly remarks, "[E]ven evidence that is to all common sense hostile to the traditional picture is rendered invisible by unjustified assumptions."[28]

Natural selection, like artificial selection, undoubtedly influences diversity up to a certain limit, but if it is the process through which every life form is designed, then we should expect to see in fossil data the various transitional stages that natural selection produced. Any fossil that looks transitional without intermediates, such as the platypus, is presumably designed that way, since information theory reveals continuity of the genome.[29] The transitional indices of evolution are not missing; they simply never existed. If biologists cannot adequately account for the fossil record, it is very unlikely that they can explain the origins of species.

Biologists can talk about how species have progressed, but they cannot explain the origin of species without knowledge of the origin of life. Leading evolutionists identified the origin of life as one of the major gaps in evolutionary theory when they marked the bicentenary of Charles Darwin's birth.[30] Why are they concerned? It is because without knowledge of the origin of life, Darwin's hypothetical tree model, which relates both extinct and extant species back to the origin of life, is not plausible. Darwin's title The Origin of Species is, therefore, a misnomer. Although empirical science rules out the "spontaneous generation of life," Darwinists, unwilling to accept the fact that God created the cosmos, choose to believe in the spontaneous generation of life as a philosophical necessity. In other words, from the outset evolutionists chose a scientifically wrong hypothesis as the foundation of their paradigm. Evolutionists can come up with circumstantial evidence, but once the foundation of a problem is wrong, every other explanation that follows is suspect. Productive science is built only on empirical facts.

Recent advances in information theory indicate that the origin of life is not knowable and, hence, is unsolvable as a scientific problem. Science is unable to explain how information began to govern chemical reactions through the means of a code.[31] The doctrine of life evolving from non-life (abiogenesis) is, therefore, a specious delusion. Science cannot prove or disprove God's existence; it also cannot explain the origin of life, since God claims to be both the Beginning and the End. This implies that the origin and end of life are outside the realm of science. Evolutionism, consequently, has no proper scientific foundation. Biologists, defending their teaching of evolution, appeal to the need for a unifying theory, similar to that of atomic theory in chemistry. But without a solid foundation can the theory of evolution be deemed unifying and reliable?

The point that Darwinian evolution is without a solid scientific foundation is not new. British biologist L. Harrison Matthews, in his introduction to a 1971 edition of *The Origin of Species*, asserts that belief in evolution parallels belief in special creation, since both are concepts that so far are incapable of being proven.[32] Are modern scientists honest, then, when they present evolutionism as a scientific fact? One must distinguish between science and scientism, which is the assumption that science is the only reliable guide to truth.[33] The Scriptures consistently maintain that God is the totality of truth.

My ninth chapter attempts to explain God's existence from three points of view. The first refutes Dawkins' argument in his best-selling book *The God Delusion*. The second section presents Antony Flew's case supporting God's existence. Until quite recently Flew was the world's most famous atheist who was far more involved than Dawkins in spreading the tenets of atheism. Flew describes his change of conviction as follows:

> I must say again that the journey to my discovery of the divine has thus far been a pilgrimage of reason. I have followed the argument where it has led me. And it has led me to accept the existence of a self-existent, immutable, immaterial, omnipotent, and omniscient Being.[34]

The third point of view presented in Chapter 9 is my personal attestation of God's existence.

E. Exploitation of Media Support

The fifth approach is the use of the media to promote evolutionism. Celebrities in evolution science, based on unsubstantiated assumptions, provide specious explanations of the origins of species in books, speeches, and interviews. The media tend to honor their views as scientific facts and disregard other possible explanations, dismissing any opposition as religiously motivated. For instance, many in the media accept Richard Dawkins' atheistic fantasies as scientific facts and disparage attempts by other accomplished scientists to point out the insurmountable problems of the evolutionary paradigm. While doing a good job in shielding the public from theistic fundamentalism, the media are inclined to celebrate atheistic fundamentalism under the banner of evolutionism. The media's influence on the controversy regarding creationism and evolutionism is the focus of my tenth chapter.

Compromising Scientific Integrity

Some may wonder what the present level of discourse has to do with scientific integrity. The recent protest by the Union of Concerned Scientists[35] indicates that the issue of scientific integrity is at stake in this debate. Biologist Ariel A. Roth reminds us in *Origin: Linking Science and Scripture* that, although the scientific enterprise is basically honest, we cannot overlook the "intellectual phase locking" or self-deception that potentiates honest mistakes.[36] Roth points to four problems that the scientific process encounters. The first is that a number of areas of reality lie outside the domain of science. The second is that historical science, unlike physical science, is not easily tested. The third is that scientists become emotionally involved in studies that culminate in philosophical conclusions. The fourth is "paradigm acceptance." For instance, "Scientists once believed in the spontaneous generation of life. Then they rejected the idea, only to later reaccept it."[37] NASIM, however, contends: "The arguments of creationists reverse the scientific process. They begin with an explanation that they are unwilling to alter."[38] *Here we see that evolutionists doing exactly the same thing.*

These problems are all connected to studies of life's origin, and the blame rests squarely on modern scientists' *a priori* belief in the doctrine of naturalism and materialism. The same evidence that led Sir Isaac Newton and Albert Einstein to believe in creation by a superior spirit prompts many modern scientists to advocate evolution by natural selection. To be true to the integrity of their discipline, scientists must not blend philosophical preference with science and must follow the evidence wherever it leads. It is intellectual fraud for a leading scientist like Dawkins to use his professional standing to champion atheism.

Reaction to Dawkins' *The God Delusion* has been mixed, but several rebuttals have appeared. Alister McGrath and Joanna Collicutt McGrath's *The Dawkins Delusion* (2007) is endorsed on the jacket by prominent atheist Michael Ruse, who as Professor of Philosophy and author of *Darwinism and Its Discontent* comments that "*The God Delusion* makes me embarrassed to be an atheist, and the McGraths show why." Other noteworthy books include Thomas Crean's *God Is No Delusion: A Refutation* of *Richard Dawkins* (2007), David Berlinski's *The Devil's Delusion: Atheism and Its Scientific Pretensions* (2008), Scott Hahn and Benjamin Wiker's *Answering the New Atheism: Dismantling Dawkins' Case Against God* (2008), and Kenneth Lawrence's *The Evolution Delusion* (2008). The present book titled *The Darwinian Delusion: The Scientific Myth of Evolutionism* differs from these studies in that it is written from the standpoints of history, religion, science, and personal experience of the divine.

I respect biologists generally for their contributions to unfolding the mysteries of our phenomenal world. As a pseudoscientific religion, however, the dogmatic creed of evolutionism, like any other dogmatism, must be prohibited from science classes. It is unacceptable that evolutionists should exploit their positions of academic authority to beguile the public, and especially children, into accepting atheism as a scientific fact. In light of this abuse of scientific knowledge, I sometimes am compelled to use simple examples and terminology that promote a wider understanding of the issues addressed. My goal is to challenge pseudoscience and the prevailing culture of scientism.

NOTES

1. *There Is a God: How the World's Most Notorious Atheist Changed His Mind* (New York: Harper, 2007), p. 183.
2. *Did Darwin Get It Right?* (Huntington, Indiana: Our Sunday Visitor, 1998), p. 42.
3. *Science, Evolution, and Creationism* (Washington, D.C.: National Academies Press, 2008), p. 24.
4. Evolution: A Theory in Crisis (Bethesda, MD: Adler & Adler, 1986), p. 358.
5. *Science, Evolution, and Creationism,* p. 24.
6. *Evolution as a Religion* (London: Methuen, 1985), p. 1.
7. "Social Responsibility and the Scientist in Modern Western Society," *Perspectives in Biology and Medicine* 14 (1971), p. 368.
8. "Origins of Species," *NewScientist,* 14 May 1981, p. 452.
9. See Richard Dawkins, *The Selfish Gene* (Oxford: Oxford University Press, 2006), p. 20.
10. God: The Failed Hypothesis. How Science Shows That God Does Not Exist (New York: Prometheus Books, 2007), p. 56.
11. Ibid, p. 131.
12. See Dawkins, *The God Delusion* (Boston: Houghton Mifflin, 2006), p. 13.
13. Quoted in Max Jammer, *Einstein and Religion: Physics and Theology* (Princeton: Princeton University Press, 1999), p. 93.
14. *The Ghost in the Universe* (New York: Prometheus Books, 2002), p. 17.
15. Ibid.
16. *The Blind Watchmaker* (New York: Penguin, 2006), p. 139.
17. Quoted in George J. Marlin, Richard P. Rabatin, and John L. Swan (eds.), *The Quotable Chesterton: A Topical Compilation of the Wit, Wisdom and Satire of G. K. Chesterton* (Sydney: Image Books, 1987), p. 113.
18. *The Case for a Creator: A Journalist Investigates Scientific Evidence That Points toward God* (Grand Rapids: Zondervan, 2004), p. 86.

19. *The Music of Life: Biology Beyond the Genome* (Oxford: Oxford University Press, 2006), p. xi.
20. "The Shape of Evolution: A Comparison of Real and Random Clades," *Paleobiology* 3.1 (1977), pp. 34-35.
21. See *http://www.pbs.org/wgbh/evolution/library/glossary/glossary.html*.
22. Ibid.
23. "How Evolution Became a Religion: Creationists Correct?" *National Post*, 13 May 2000, B1.
24. See Mark Ridley, *The Problems of Evolution* (Oxford: Oxford University Press, 1985), p. 4.
25. *What Evolution Is* (New York: Basic Books, 2001), p. 163.
26. *The God Delusion*, p. 73.
27. Ibid, p. 19.
28. Evolution: A Theory in Crisis, p. 353.
29. "Scientific Reality vs. Intelligent Design's False Claims," *http://www.cynthiayockey.com/pages/1/index.htm*
30. "Evolution's Final Frontiers: 16 Challenges for a Modern-Day Darwin," *NewScientist*, 31 January-6 February 2009, pp. 41-43.
31. "Scientific Reality vs. Intelligent Design's False Claims," *http://www.cynthiayockey.com/pages/1/index.htm*
32. Introduction, *The Origin of Species* (London: J. M. Dent & Sons, 1971), pp. x-xi.
33. See J. F. Haught, *Science and Religion: From Conflict to Conversation* (New York: Paulist Press, 1995), p. 16.
34. There Is a God, p. 155.
35. "Restoring Scientific Integrity in Policymaking," *http://www.ucsusa.org/scientific_integrity/interference/scientists-signon-*statement.html. Retrieved on 30 June 2008.
36. *Origins: Linking Science and Scripture* (Hagerstown, MD: Herald Publishing, 1998), p. 294.
37. Ibid, pp. 294-95.
38. *Science, Evolution, and Creationism*, p. 43.

CHAPTER 1

EVOLUTIONISM:
PURPOSE AND CONCERNS

Science is not the only way of knowing and understanding . . . . *But science is a way of knowing that differs from other ways in its dependence on empirical evidence and testable explanations.* Because biological evolution accounts for events that are also central concerns of religion—including the origins of biological diversity and especially the origins of humans—evolution has been a contentious idea within society since it was first articulated by Charles Darwin and Alfred Russel Wallace in 1858.[1]

—NASIM

Science knows about objective reality, the mask of matter that our five senses detect. But the mind goes beyond the five senses. And what lies beyond the boundary of the five senses holds enormous mysteries, and it does Dawkins no good to lump the two worlds of inner and outer together.[2]

—Deepak Chopra

Evolutionism is perhaps the most jealously guarded dogma of the American public philosophy. Any sign of serious resistance to it has encountered fierce hostility in the past, and it will not be abandoned without a tremendous fight. The gold standard could go (glad to be rid of that!), Saigon abandoned, the Constitution itself slyly junked. But Darwinism will be defended to the bitter end.[3]

—Tom Bethell

WHAT EVOLUTIONISM IS

In general usage, evolution simply means "change over time." In this sense evolution is a fact, and we do not need any theory to express it. However, when we ask "change over time from what to what?" we need experiential knowledge and common sense to know when evolution is a fact and when it is not. Under these circumstances we require a theory to describe the mechanisms involved and their limitations. Evolutionism is the belief in the Darwinian paradigm of evolution beyond the limits of scientific verification. Outside the scientific domain the evolutionary worldview is tainted with philosophical assumptions and conclusions as it addresses the mystery of the origin of life and the universe. To understand why evolutionism, the antithesis of creationism, is a concern, we need to unfold its meaning, purpose, and dynamics.

The molecules-to-molecules or bacteria-to-bacteria framework of evolution falls within the limits of science and, hence, constitutes genuine evolutionary science. I refer to bacteria because they are the most rudimentary living organisms. At this level evolution "produces relatively small-scale *microevolutionary* changes in organisms."[4] NASIM defines microevolution as "changes in the traits of a group of organisms that do not result in a new species."[5] The scientific community is unanimous in its judgment that microevolution is a scientific fact. This is the type of evolution manifest in laboratory studies used to combat diseases and aid in agricultural diversity. According to NASIM, "to be accepted, scientific knowledge has to withstand the scrutiny of testing, retesting, and experimentation."[6] The bacteria-to-human version of evolution that constitutes evolutionism is merely a scientific myth because it falls short of the requirements for testing, retesting, and experimentation. This brand of evolution (macroevolution) is notorious for its religious and social implications.

To give macroevolution a patina of scientific status, proponents try to relate it to the microevolution. Accordingly, NASIM claims that "incremental evolutionary changes can, over what are usually very long periods of time, give rise to new types of organisms, including new species."[7] Evolutionism is, therefore, the extrapolation from bacteria-to-bacteria evolution (variations *within* species) to bacteria-to-human evolution (variation *between* species). In the former model no new organisms are formed; consequently, origin is not important, and time is not an issue. Testing, retesting, and experimentations can be accomplished. In bacteria-to-human evolution, however, new organisms are imagined but not observed. As such, the length of requisite time and the emergence of various transitional organisms cannot be subject to testing, retesting, and experimentation. Macroevolution is, therefore, unscientific.

From a philosophical point of view, the *origin of species* and the *diversity of species* are two different matters. The origin of species has religious implications, but the diversity of species does not. We can address the diversity of species without a prior understanding of the origin of life. However, to address correctly the origin of species, we must begin with the origin of life. The life history of species will differ depending on whether life arose spontaneously from dead matter or was created. For instance, if life evolved from dead matter, it would be mindless and devoid of spiritual values. If the origin of life is unknowable, then, the origin of species is undecidable. Scientists are wrong when they treat the origin of life and the origin of species as unrelated. Douglas Theobald, for example, disconnects the origin of life from the *origin of species* in the following argument:

> In evolutionary theory it is taken as axiomatic that an
> original self-replicating life form existed in the distant

past, regardless of its origin. All scientific theories have their respective, specific explanatory domains; no scientific theory proposes to explain everything. Quantum mechanics does not explain the ultimate origin of particles and energy, even though nothing in that theory could work without particles and energy. Neither Newton's theory of universal gravitation nor the general theory of relativity attempt[s] to explain the origin of matter or gravity, even though both theories would be meaningless without the *a priori* existence of gravity and matter.[8]

Theobald's argument would be invalid if we replaced "quantum mechanics" with "the origin of quantum mechanics," or "theory of universal gravitation" with "theory of the origin of universal gravitation," or "general theory of relativity" with "general theory of the origin of relativity." When the term "origin" is included in any of these descriptions, it automatically demands explanations. Had Darwin titled his book *Species* or *Species by Common Descent* as opposed to *The Origin of Species*, his seminal text would not have generated so profound an impact. The word "origin" takes science into a religious/philosophical domain. We can look at this point from another angle. If we focus on the contents of this book, knowledge of the author is irrelevant. But when we consider the origin of this book, then knowledge of the author, the mind that produced it, is relevant. Let me clarify this point.

The life of this book is the information (non-matter) expressed in printed form on paper (matter). Without the information this book fails to exist or has no life. This book, therefore, consists of a material and non-material component. We cannot bypass the origin of the information in this book and hope to explain the book's origin. The same argument applies to the origin of species. A species consists of a material component (body) and non-material component (life); hence, it is analogous to a book. The life that dwells within the body is analogous to the information printed on its pages. The body of a species is dependent on life for its existence. To describe the origin of a species, we must first address the origin of life. We cannot bypass the origin of life (the independent component) and hope to explain the origin of species (the dependent component). If the origin of life is unknowable, it is foolhardy to imagine that the origin of species is knowable. By analyzing the contents (material) of this book, scientists cannot establish its origin; similarly, by examining the material composition of a species, scientist cannot establish its origin. If the origin of life is not a scientific problem, the same is true of the origin of species. The origin of life or species as a scientific problem cannot be tested or repeated; in this sense science is not based on repeated facts but on human speculation. Hence, under the Darwinian paradigm of bacteria-to-human evolution, science is not a way of knowing that differs from other ways such as religion or common sense.

As a further illustration, without my name printed on this book, its origin is not knowable through science. However, common sense indicates that the information and, hence, the book must have originated from a mind and was not generated by chance. For instance, it is foolish to imagine that a mindless process can create organisms with abstract minds; similarly, it is preposterous to postulate that robots can create human beings. Therefore, the origin and diversity of species will be remarkably different based on whether life evolved by chance or was created. Without knowledge of the independent variable (life), there is no scientific foundation for Darwin to theorize that the dependent variable (species) evolved in a manner that resembles the structure of a tree.

Modern evolutionists, conscious of the fragile basis of the Darwinian paradigm, are defensive about his theory. However, confronting roadblocks in their research studies, some evolutionists have conceded publicly that Darwin was wrong. Graham Lawton, the features editor of *NewScientist*, thus writes:

> Ever since Darwin the tree has been the unifying principle
> for understanding the history of life on Earth. At its base
> is LUCA, the Last Universal Common Ancestor of all living
> things, and out of LUCA grows a trunk, which splits again
> and again to create a vast, bifurcating tree. Each branch

represents a single species; branching points are where one species become two. Most branches eventually come to a dead end as species go extinct, but some reach right to the top—these are living species. The tree is thus a record of how every species that ever lived is related to all others right back to the origin of life.[9]

The above is a concise description of the tree of life Darwin envisioned, which scientists have been trying to validate for some time:

> For much of the past 150 years, biology has largely concerned itself with filling in the details of the tree. "For a long time the holy grail was to build a tree of life," says Eric Bapteste, an evolutionary biologist at the Pierre and Marie Curie University in Paris, France. A few years ago it looked as though the grail was within reach. But today the project lies in tatters, torn to pieces by an onslaught of negative evidence. Many biologists now argue that the tree concept is obsolete and needs to be discarded. "We have no evidence at all that the tree of life is a reality," says Bapteste. That bombshell has even persuaded some that our fundamental view of biology needs to change.[10]

The truth, of course, is that the origin of life must be known in order to predict correctly the origin of species. If consistency of natural processes is the rule in science, then scientists know that life comes only from preexisting life. This truth is empirically observable in our daily experience. Accordingly, the origin of all life must be a *living being*; the source of all minds, therefore, must be a *living mind*. This conclusion, however, is anathema to scientists, especially those of atheistic backgrounds. However, because science is based on facts, not theoretical preference, irrespective of what modern scientists advocate, the whole charade about evolutionism and creationism hinges on the origin of life. If life is the product of creation, then a purely evolutionary worldview is without foundation. Because they resist this inference, many scientists choose instead to believe that life originates from non-life.

By accepting the wrong premise, such scientists are desperate to demonstrate how life originates from non-life. In order to achieve this objective, the scientific community has set up "The Origin-of-Life Science Foundation" under the following terms:

a) The Origin-of-Life Science Foundation should not be confused with "creation science" or "intelligent design" groups. It has no religious affiliations of any kind, nor are we connected in any way with any New Age, Gaia, or "Science and Spirit" groups. The Origin-of-Life Science Foundation, Inc. is a science and education foundation encouraging the pursuit of natural-process explanations and mechanisms within nature. The Foundation's main thrust is to encourage interdisciplinary, multi-institutional research projects by theoretical biophysicists and origin-of-life researchers specifically into the origin of genetic information/instructions/message/recipe in living organisms. By what mechanism did *initial* genetic code arise in nature?[11]

b) "The Origin-of-Life Prize" ® (hereafter called "the Prize") will be awarded for proposing a highly plausible *mechanism* for the spontaneous rise of *genetic instructions* in nature sufficient to give rise to life. To win, the explanation must be consistent with empirical biochemical, kinetic, and thermodynamic concepts as further delineated herein, and be published in a well-respected, peer-reviewed science journal(s).[12]

c) The one-time Prize will be paid to the winner(s) as a twenty-year annuity in hopes of discouraging theorists' immediate retirement from productive careers. The annuity consists of $50,000.00 (U.S.) per year for twenty consecutive years, totaling one million dollars in payments.[13]

d) Other than announcements in scientific journals, The Prize will not be publicly advertised in lay media. The Origin-of-Life Science Foundation, Inc. wishes to keep the project as quiet as possible *within the scientific community*.

No media interviews will be granted until after the Prize is won.[14] (Emphasis theirs.)

It is now over a decade since the Foundation's inception. Instead of announcing a winner, the program was temporarily placed on hold in 2008.[15] Worse still, NASIM has asserted, "Even if a living cell could be made in the laboratory from simpler chemicals, it would not prove that nature followed the same pathway billions of years ago on the early Earth."[16] This means that we will never know how life originated on Earth. World-famous biophysicist Hubert P. Yockey confirms this fact. Yockey is the leading information theorist whose work is broadly acknowledged by The Origin-of-Life Science Foundation.[17] Yockey stipulates that "[t]he origin of life, like the origin of the universe, is unknowable."[18] Emilio Segre, winner of the Nobel prize for Physics in 1959, contended earlier that the origin of the universe is not a scientific question since scientific theories are validated by experiment, consistency tests, and predictive power, all of which are hardly applicable to the universe's origin.[19]

Since the origin of life is unsolvable as a scientific problem, it follows that scientists are not qualified by virtue of their discipline to reach valid conclusions about such philosophical issues. Accordingly, the title *The Origin of Species* that Darwin used to describe his "finches-to-finches" concept of evolution is inaccurate. The theory simply expresses the diversity of species; it says nothing about the origin of species.

Darwin discusses "origin" primarily in Chapter 1's section on "Variations Under Domestication" with four different subheadings. For instance, under the first subheading, "Character of Domestic Varieties; Difficulty of Distinguishing between Varieties and Species; Origin of Domestic Varieties from One or More Species," the word "origin" appears three times in the six paragraphs that follow:

- The argument mainly relied on by those who believe in the multiple *origin* of our domestic animal is that we find in most ancient times, on the monuments of Egypt, and in the lake-habitations of Switzerland, much diversity in the breeds; and that some of these ancient breeds closely resemble, or are even identical with, those still existing.

- The *origin* of most of our domesticated animals will probably forever remain vague.
- The doctrine of the *origin* of our several domestic races from several aboriginal stocks has been carried to an absurd extreme by some authors . . . . We must admit that many domestic breeds must have originated in Europe; for whence otherwise could they have been derived?[20] (Emphasis mine.)

Here the word "origin" is used in the sense of "species' country of origin." Darwin makes no contribution to science in the above quotations.

Under the second subheading, "Breeds of the Domestic Pigeon, Their Differences and Origin," the word "origin" appears only once:

I have discussed the probable *origin* of domestic pigeons at some, yet quite insufficient, length; because when I first kept pigeons and watched the several kinds, well knowing how truly they breed, I felt fully as much difficulty in believing that since they had been domesticated they had all proceeded from a common parent, as any naturalist could in coming to a similar conclusion in regard to the many species of finches, or other groups of birds, in nature.[21] (Emphasis mine.)

Here the word "origin" is used to express the different kinds of pigeons produced by artificial selection. This is merely an example of pigeon-to-pigeon microevolution. Darwin postulates "vertical" descent, but scientists are limited to "horizontal" descent. Darwin's tree of life is, therefore, a myth.

Under the third subheading, "Methodical and Unconscious Selection," the word "origin" appears twice:

> These views appear to explain what has sometimes been
> noticed—namely, that we know hardly anything about
> the *origin* or history of any of our domestic breeds. But,
> in fact, a breed, like a dialect of a language, can hardly
> be said to have a distinct *origin*.[22] (Emphasis mine.)

Here Darwin confesses that he has no knowledge of domestic
breeds' origin. Again, the title of his book is misleading.

Under the fourth subheading, "Circumstances Favourable to Man's
Power of Selection," the word "origin" is used a few times:

> To sum up on the *origin* of our domestic races of
> animals and plants, changed conditions of life are of
> the highest importance in causing variability, both by
> acting directly on the organization, and indirectly by
> affecting the reproductive system . . . . In some cases
> the intercrossing of aboriginally distinct species appears
> to have played an important part in the *origin* of our
> breeds.[23] (Emphasis mine.)

Here the use of "origin" describes only the production of
different breeds, not the origin of species.

In Chapter 6, "Difficulties of the Theory," Darwin includes a section
subtitled "On the Origin and Transitions of Organic Beings with
Peculiar Habits and Structure," but ironically he makes no reference
to "origin" throughout the entire six-page section. In Chapter 9, "On
Hybridism," one again expects Darwin to address the issue under
a section titled "Origin and Causes of the Sterility of First Crosses
and of Hybrids," but the term "origin" nowhere appears.

In Chapter 15, "Recapitulation and Conclusion," the phrase
"natural selection" is mentioned twenty-eight times and the word
"origin" just thrice. The statements about "origin" are as follows:

- It is no valid objection that science as yet throws no light on
  the far higher problem of the essence or *origin* of life.
- When the views advanced by me in this volume, and by Mr.
  Wallace, or when analogous views on the *origin* of species

are generally admitted, we can dimly foresee that there will be a considerable revolution in natural history.

- Much light will be thrown on the *origin* of man and his history. [24] (Emphasis mine.)

The views that Darwin advanced tell us nothing concrete about the origin of anything. And if science sheds no light on the problem of the origin of life, it cannot shed light on the origin of species, which depends on the origin of life. The irrelevance of the term "origin" in Darwin's famous text is reflected in his recapitulation and concluding remarks. It is also noteworthy that the book's index contains the terms "natural selection" and "varieties" but not "origin."

Darwin's overall deployment of the word "origin" is reminiscent of such usage by immigration or police officers in finding out people's country of origin. Indeed, his use of the word is plagued with so much uncertainty that it is simply misleading. Little wonder, then, that physicist H. S. Lipson remarks:

- Darwin's book—*The Origin of Species*—I find quite unsatisfactory: it says nothing about the origin of *species*; it is written very tentatively, with a special chapter on "Difficulties on Theory"; and it includes a great deal of discussion on why evidence for natural selection does not exist in the fossil record. Darwin, I think, has been ill-served by the strength of his supporters.
- An article that I published in *Physics Bulletin* (May 1980, p. 138), stating my views, has shown me that many people have doubts like my own.
- It seems to me that, in our present state of knowledge, creation is the only answer—but not the crude creation envisaged in Genesis.
- As a scientist, I am not happy with these ideas. But I find it distasteful for scientists to reject a theory because it does not fit in with their preconceived ideas. [25]

Nobel laureate Arno Penzias contends that creation is supported by all the data so far. [26] Lipson is right that Darwin "has been ill-served by the strength of his supporters." It is my view, therefore, that Darwin did not present a theory on either the origin of life or the origin of species. In his book *Creation: Facts of Life*, ex-evolutionist Gary Parker affirms this point: "In spite of the title of his book, *The Origin of Species,* the one thing Darwin never really dealt with was the *origin* of species."[27]

Science curricula can address topics such as the diversity of species without resorting to the word "origin," which relates to the domains of philosophy and religion. In light of this point L. Harrison Matthews, in his introduction to a 1971 edition of *The Origin of Species*, writes:

> In accepting evolution as a fact, how many biologists pause to reflect that science is built upon theories that have been proved by experiment to be correct, or remember that the theory of animal evolution has never been thus proved? . . . The fact of evolution is the backbone of biology, and biology is thus in the peculiar position of being a science founded on an unproved theory—is it then a science or a faith? Belief in the theory of evolution is thus exactly parallel to belief in special creation—both are concepts which believers know to be true but neither, up to the present, has been capable of proof.[28]

Belief in the theory of evolution is as religious as belief in special creation by God. To argue that Darwinian evolution is science makes a mockery of the scientific enterprise.

Two essentially religious views are at war within the scientific community. The first, evolutionism, is able to hide under the banner of science because it relies on the god of chance and materialism; the other, creationism, cannot hide because it derives from both science and sacred Scriptures. It is therefore important to clarify certain points about religion and science.

Evolution encompasses two autonomous fields: science (microevolution) and a secular worldview (evolutionism). In the religious world the controversy is between theism and atheism. In the scientific world the controversy concerns creationism versus evolutionism as divergent beliefs. *There is thus no "creation-evolution" controversy.* Evolutionists insinuate that the debate concerns creationism versus evolution. Accordingly, NASIM's 2008 booklet is titled *Science, Evolution, and Creationism* instead of *Science, Evolutionism, and Creationism.* In this way the public is led to believe that the war is between scientists and religious groups. However, most religious thinkers accept some degree of evolution, a fact that NASIM and the Council of Europe Parliamentary Assembly confirm. NASIM thus states, "Many religious people accept the reality of evolution, and many religious denominations have issued emphatic statements reflecting this acceptance."[29] The battle, in other words, is within the scientific community. The creationism-evolutionism controversy is a struggle between scientists who want to replace the role of God by natural selection and those who accept the premise of divine creation. We shall next examine how evolutionism has infected our educational and social systems.

THE CONSEQUENCES OF EVOLUTIONISM

Indoctrination under the Disguise of Science

Except for Dawkins, who publicly maintains that evolutionism leads to atheism, few leading scientists are transparent regarding their philosophical presuppositions for fear of losing public funding. It is educationally, socially, and politically incorrect for scientists to draw conclusions about the beginning and end of life. Many in the public domain cannot distinguish between dogma and scientific fact; anything that is presented under the aegis of science tends to be accepted as apodictic truth.

Citizens at large should be concerned with two points in science education. One is the separation of science and religion; the other is the indoctrination of children. These points are stressed in the draft resolution of the Council of Europe Parliamentary Assembly:

> There is a real risk of a serious confusion being introduced
> into our children's minds between what has to do with

convictions, beliefs and ideals and what has to do with science, and of the advent of an "all things are equal" attitude, which may seem appealing and tolerant but is actually disastrous.[30]

The tenets of evolutionism, however, are antithetical to Christian faith. Creationism professes the law of love; evolutionism endorses survival of the fittest. Creationism posits the idea of life after death; evolutionism believes in no afterlife. Evolutionism thus imposes its set of religious views on students as scientific "facts." Fairness demands that students know of the opposing views as well.

Why is it that many scientists, particularly Americans, cannot muster the courage to tell the media and courts that evolutionism leads to atheism? In this regard Dawkins is honest for not wanting to testify in court. During his visit to Toronto in the summer of 2007, Dawkins said:

> I'm not a good politician . . . . One of the things the creationist lobby wants to hear is that evolutionism leads to atheism, and since I'd have to say that to a jury, the evolutionist would lose the case immediately.[31]

When court proceedings focus on what constitutes science versus religion, little or no attention is given to the far more important point of student indoctrination. In *The God Delusion* Dawkins deals with what he claims to be the religious indoctrination of children. He argues that it is nothing less than "child abuse" to label children as Protestants, Catholics, Muslims, and Jews, but, in applauding his colleagues for winning court cases so that children can be exposed only to the doctrines of evolutionism, he is guilty of the same travesty.

If Dawkins can readily advocate the tenets of evolutionism in the public arena, one can only imagine what takes place in his science classes. Of course, Dawkins will not let the world second-guess how far he will go to promote evolutionism. In his book titled *The Blind Watchmaker*, he merrily shares one of his presumably many victories in championing evolutionism. He reports his conversion of an American creationist student as follows:

I was reminded of the creationist student who, through some accident of the selection procedure, was once admitted to the Zoology Department at Oxford University. He had been educated at a small fundamentalist college in the United States and had emerged a simple, young-Earth creationist. When he arrived in Oxford, he was encouraged to attend a course of lectures on evolution. At the end he came up to the lecturer (who happened to be me), beaming with the primal joy of discovery: 'Gee', he exulted, 'this evolution! It really makes sense.' It certainly does. In the words of a tee-shirt which an anonymous American reader was kind enough to send me: 'Evolution—The Greatest Show on the Earth—The Only Game in Town!'[32]

This anecdote suggests that Dawkins uses his university classroom as a recruiting ground to proselytize students. Here is an affirmation that evolutionism is a full-fledged alternative to biblical creationism. The educational, political, and legal understanding of the separation between religion and state in the Western world seems to be limited to creationism. Is the evolutionist community at large, however, not a congregation by any other name?

We get to know more of the motives that drive evolutionists to resist creationism from those who at some point reverse their conviction. For example, Gary Parker, makes this admission in *Creation: Facts of Life*:

> For me, "evolution" was much more than just a scientific theory. It was a total world-and-life view, an alternate religion, a substitute for God. It gave me a feeling of my place in the universe, and a sense of my relationship to others, to society, and to the world of nature that had ultimately given me life. I knew where I came from and where I was going . . . . I didn't just believe evolution; I embraced it enthusiastically! And I taught it enthusiastically. I considered it one of my major missions

as a science professor to help my students rid themselves completely of old, "pre-scientific" superstitions, such as Christianity. In fact, I was almost fired once for teaching evolution so vigorously that I had Christian students crying in my class![33]

We see here a representative pattern. The objective of many scientists is to promote evolutionism under the rubric of "Evolution Science" while derogating Christianity. The covert goal of evolutionism is nothing less than to form an agnostic and atheistic population. Meanwhile politicians, the media, and curriculum planners appear to be ignorant of the intellectual threat posed by evolutionism.

The Council of Europe Parliamentary Assembly's draft resolution addresses "the dangers of creationism in education" but fails to realize the danger of evolutionism. In order to convince the public of the dangers of creationism, the Council should provide case studies of how creationism taught alongside evolutionism has eroded an understanding of science. Educators, politicians, and judges seem to ignore the relevance of what transpires in actual learning environments. Not only the curriculum matters but also the method of delivery. Physicist J. Willits Lane, in a letter to the journal *Physics Today*, describes the disdainful atmosphere he experienced in science classes that promote evolutionism:

After reading a spate of virulently anti-creationist articles and letters in your publication, I decided that something less virulent and more thoughtful should be said . . . .

As we might all easily agree, it isn't very scientific to make assumptions dogmatically and then accept only evidence in favour of these assumptions. It is the practice of this precept that separates the unbelievers from the believers, sheep from goats, and so forth. Most of us, history says, will test as goats. Therefore, a word of caution: How much do we actually know (other than that it has something to do with someone's religion) about this set of ideas we are calling "creationism"? I shall confess that I know nothing. Will any of the noisemakers out there also confess?

I do know what we do not know about creation: almost everything. Science, like religion, is not a physical thing itself, but a non-material set of ideas. It is an ideology and is not exempt from the scrutiny to which we subject other ideologies . . .

We have several things to gain by lowering our voices. One is the possibility that paying attention to some radically different ideas, however wacky, may suggest to us an insight into science that we do not expect. For instance, we do not have a thoroughly rational, tested hypothesis about the origin of our species. Indeed, we haven't even been able to agree upon a biological classification system for primates. Somewhere buried in the creationist arguments may be the right question, one that we have been ignoring because it wasn't proper to consider it! The second thing we have to gain is our decency and humanity. I have myself sat in class after class in sciences and humanities in which any idea remotely religious was belittled, attacked, and shouted down in the most unscientific and emotionally cruel way. I have seen young students raised according to fundamentalist doctrine treated like loathsome alley cats, emotionally torn apart, and I never thought that this sort of treatment was any better than the treatment that religious prelates, who held authority, gave Galileo. Why scream about the inhumanity of nuclear war if you are also willing to force people of fundamentalist faiths to attend public schools in which their most cherished beliefs will be systematically held up to ridicule and the young children with it? These people are mostly too poor for private schools to be an alternative. The state tries to prevent them from teaching their children at home rather than sending them to school. What choices do they have? Would you call it freedom? Do you call it fair?

> Is it really a terrible thing for a textbook to mention that, aside from the Darwin theory of evolution, there have existed other ideas, many of them religious in nature? Would that not open the mind of students rather than close them to scientific possibilities? Wouldn't it make the fundamentalist student feel a little more welcome and better equip him to take an unbiased view of evolution?[34]

To limit students' scientific reasoning to the framework of Darwin's theory discourages free thinking and fosters censored learning. Fear of comparison with the opposing worldview shows incompetence rather than confidence in a field that is deemed to be as securely established as other natural sciences.

The Committee on Culture, Science, and Education that prepared the draft resolution for the Council of Europe Parliamentary Assembly does not seem to understand the difference between evolution (a scientific process) and evolutionism (a scientific belief). For instance, the "spontaneous rise of *genetic instructions* in nature sufficient to give rise to life" is a belief that is to be justified by developing a "highly plausible mechanism." I dispute this postulate, for it is not a scientific fact. Life derives from preexisting life (biogenesis), and this truth is an indisputable scientific law that negates the evolutionist precept of the "origin of species based exclusively on evolution by natural selection." *The key word again is "origin," which marks the difference between "evolution" and "evolutionism."* As is evident in the following quotation from its draft resolution, the Council seems oblivious of this core distinction:

> Creationists question the scientific character of certain items of knowledge and argue that the theory of evolution is only one interpretation among others. They accuse scientists of not providing enough evidence to establish the theory of evolution as scientifically valid. On the contrary, they defend their own statements as scientific. None of this stands up to objective analysis.[35]

As a point of correction, most creationists are accomplished scientists who question evolutionists' knowledge of the origin of life. This knowledge, as the scientific community is fully aware, is required for understanding the origin of species. Physicists and chemists teach the nature of atoms without reference to the *origin* of atoms. Biologists can effectively teach the nature of organisms/species without reference to the *origin* of organisms/species. Without reference to origin, they use evolution to explain how organisms/species adapt to their environment by natural selection, how different organisms/species share characteristics, and how the diversity of life came to be by speciation.

The Council of Europe Parliamentary Assembly also states:

> The teaching of all phenomena concerning evolution as a fundamental scientific theory is therefore crucial to the future of our societies and democracies. For that reason it must occupy a central position in the curriculum, and especially in the science syllabus. Evolution is present everywhere, from medical overprescription of antibiotics that encourages the emergence of resistant bacteria to agricultural overuse of pesticides that causes insect mutations on which pesticides no longer have any effect.[36]

The point that "[e]volution is present everywhere" is not in dispute. What the Council here ignores is the fact that a good number of creationists are accomplished scientists who believe in God as the Creator, as opposed to evolutionists who believe in natural selection as the designing instrumentality. The medical and agricultural applications of evolution fall within the scientific domain and are accepted by both creationists and evolutionists. It is good science. Evolution by natural selection, however, assumes that human beings and simians have an unknown common ancestor. It is a belief that cannot be presented as a scientific fact. If it were true, evolutionists would be able to provide a laboratory demonstration of an organism that is a probable intermediary between two different organisms.

The Council yet endorses evolutionism: "We are witnessing a growth of modes of thought which, the better to impose religious dogma, are attacking the very core of the knowledge that we have patiently built up on nature, evolution, our origins and our place in the universe."[37] The subject of human origins is not a scientific question. A worldview, whether religious or scientific, must not be imposed on others as a fact without being subject to debate and scholarly examination. This undemocratic approach compromises scientific integrity.

Ditching Scientific Integrity

The integrity of the modern scientific community, particularly in the Western world, is challenged both from within and from without. Opposition to the Darwinian theory is strongest within that community itself. Once a community's integrity is questioned in one area, it becomes suspect in others. Significantly, the Union of Concerned Scientists today is seeking the signed support of scientists to restore integrity in policymaking.[38] Among its list of concerns is "Science, Evolution, and Intelligent Design."

In the following paragraphs we shall learn how advocates of Darwinian evolution undermine scientific integrity in their effort to promote evolutionism. The discussion will be based on the Nobel laureate Richard Feynman's ground rule for cultivating scientific integrity. Wikipedia provides this relevant background:

> In 1974 Feynman delivered the Cal Tech commencement address on the topic of cargo[-]cult science, which has the semblance of science but is only pseudoscience due to a lack of "a kind of scientific integrity, a principle of scientific thought that corresponds to a kind of utter honesty" on the part of the scientist. He instructed the graduating class that "The first principle is that you must not fool yourself—and you are the easiest person to fool. So you have to be very careful about that. After you've not fooled yourself, it's easy not to fool other scientists. You just have to be honest in a conventional way after that.[39]

Feynman refers to "cargo-cult science" because it follows all the forms of scientific investigation but is missing something essential. He identifies this essential ingredient as "a kind of scientific integrity, a principle of scientific thought that corresponds to a kind of utter honesty—a kind of leaning over backwards."[40] He argues that it is "this type of integrity, this kind of care not to fool yourself, that is missing to a large extent in much of the research in cargo[-]cult science."[41] He contends that "ordinary people with commonsense ideas are intimidated by this pseudoscience."[42]

Let us examine neo-Darwinian evolutionism in light of Feynman's "cargo-cult" criteria for scientific integrity. We begin with a story that archaeologist and theologian Victor Pearce reports in *Evidence for Truth: Science*:

> In 1981—to mark its centenary—the British Museum opened an exhibition on Darwinism. Visitors were taken aback to see two notices at the entrance to the exhibition. Both asked about the origins of living creatures, but gave two opposite explanations.
>
> - The first notice stated: "One idea is that all living things have evolved from a distant ancestor by a process of gradual change—an explanation first thought of by Charles Darwin."
> - The second notice stated: "Another view is that God created all living things perfect and unchanging."
>
> This produced an outcry in the press, but 22 of the Museum's biologists quickly replied, "Are we to take it that evolution is a fact, proven to the limits of scientific rigour? If that is the inference, then we must disagree most strongly."[43]

Pearce's story is summarized and posted at www.believershope. com in a question-and-answer format under the title "Evolution—The Biggest Cover-Up since ROSWELL":

QUESTION: In 1981 22 British Museum biologists said, "Evolution is not a fact[,]" yet still today we are being bombarded by *evolution* as the only credible way man came to be on the earth. It is a fact that more scientists today believe evolution as a theory can no longer be taken seriously. So why are we still being force-fed this "theory"?

ANSWER: The answer is Money. There are too many people making too much money in careers dedicated to evolution. It is a science that has not evolved with evidence[;] it hides what it does not like, destroys what it fears, and invests what it needs. It has the clout to silence the truth, and anyone who dares to question.

Let us briefly examine the above quotation. While doing so, I will make frequent reference to Feynman's ground rule for cultivating scientific integrity.

♣ *Evolutionism is about money.*

In American schools the evolution enterprise is a big business for school boards, academic institutions, media, politicians, and atheists. It began when Sir Julian Huxley delivered his Darwin Centennial Convocation address on 26 November 1959 at the University of Chicago on the 100th anniversary of *The Origin of Species*. Here is an excerpt from his speech:

> In the evolutionary pattern of thought there is no longer either need or room for the supernatural. The earth was not created; it evolved. So did all the animals and plants that inhabit it, including our human selves, mind and soul as well as brain and body. So did religion . . . . Evolutionary man can no longer take refuge from his loneliness in the arms of a divinized father-figure whom

he has himself created, nor escape from the responsibility of making decisions by sheltering under the umbrella of Divine Authority, nor absolve himself from the hard task of meeting his present problems and planning his future by relying on the will of an omniscient, but unfortunately inscrutable, Providence.[44]

James Perloff elaborates on the impact of Huxley's speech:

That year the National Science Foundation, a U.S. government agency, granted $7 million to the Biological Sciences Curriculum Study (BSCS), which began producing high school biology textbooks with a strong evolutionary slant. Given taxpayer funding, market considerations were no longer a worry. In the 1960s, public schools started using BSCS textbooks. In the meantime, surviving Southern anti-evolution laws were repealed or struck down by the Supreme Court. Students of the sixties thus faced a two-edged sword. On one hand, they were taught evolution, which effectively repudiated God and the biblical version of creation. On the other hand, Supreme Court rulings prohibited teachers from discussing God, reading from the Bible, or praying. It was legal to *deny* God's existence, but *illegal* to affirm it.[45]

Attempts now to challenge evolutionism will reduce revenues from the enormous number of extant textbooks on evolution. Discussing some of these implications in *Evolution: A Theory in Crisis*, molecular biologist and physician Michael Denton remarks:

Any suggestion that there might be something seriously wrong with the Darwinian view of nature is bound to excite public attention, for if biologists cannot substantiate the fundamental claims of Darwinism, upon which rests so much of the fabric of twentieth-century thought, then the intellectual and philosophical implications are immense.[46]

Joanna Rutkowska expresses similar concerns: "Polish politicians' recent denial of the theory of evolution is very dangerous, not only because it goes against the scientific paradigm, but [also] because it weakens society's trust in scientists and in research."[17]

NASIM generates considerable revenue through selling booklets that promote evolutionism as science and creationism as religion. It does so with such unrelenting determination that it sometimes offers the public conflicting statements. Compare, for instance, the organization's following assertions in its 2008 booklet titled *Science, Evolution, and Creationism*: "[T]he claims of intelligent[-]design creationists are disproven by the findings of modern biology" and "Intelligent design is not a scientific concept because it cannot be empirically tested."[18] Creationism and evolutionism are two mutually exclusive beliefs. Neither can be empirically established as true. Life either evolved or was created; there are no other explanations. It is a case of either abiogenesis or biogenesis.

To further discredit creationism and promote evolutionism, NASIM writes:

> The arguments of creationists reverse the scientific process. They begin with an explanation that they are unwilling to alter—that supernatural forces have shaped biological or Earth systems—rejecting the basic requirements of science that hypotheses must be restricted to testable natural explanations. Their beliefs cannot be tested, modified, or rejected by scientific means and thus cannot be a part of the processes of science.[19]

Biogenesis, however, is the hypothesis of creationism. It is a scientific fact and what we know to be true from experience. The hypothesis for evolutionism, on the other hand, is abiogenesis or the spontaneous generation of life. Louis Pasteur disproved the myth that life can originate from dead and inorganic matter. Since biogenesis and abiogenesis are mutually exclusive propositions, the verification of biogenesis as a scientific law makes abiogenesis a scientific myth.

When members of a scientific community privilege a particular philosophical conclusion, it is deemed an obsession. But when such persons insist that their philosophical preference is the only possibility and strive to prove this even in the face of contradictory evidence, it then becomes a delusion. In this context delusion means "a persistent false belief in abiogenesis held in the face of strong contradictory empirical and experiential evidence of biogenesis." That culminates in the loss of scientific integrity. The offer of a million dollars to scholar/s who can propose "a highly plausible mechanism for the spontaneous rise of genetic instructions in nature sufficient to give rise to life" is additional evidence that evolutionism is big business.

♣ *Evolutionism has evolved with no evidence.*

Concerning the lack of empirical evidence for evolutionism, Theodosius Dobzhansky asserts:

> These evolutionary happenings are unique, unrepeatable, and irreversible. It is as impossible to turn a land vertebrate into a fish as it is to effect the reverse transformation. The applicability of the experimental method to the study of such unique historical processes is severely restricted before all else by the time intervals involved, which far exceed the lifetime of any human experimenter. And yet it is just such impossibility that is demanded by anti-evolutionists when they ask for "proofs" of evolution which they would magnanimously accept as satisfactory.[50]

Belief in evolution demands more faith than belief in Scripture. Evolutionism, as Dobzhansky affirms, is not subject to scientific experimentation and falsification; hence, it is outside the realm of proper science. It is an opinion that belongs to the realm of cargo-cult science, which compromises the discipline's integrity.

♣ *Evolutionism hides what it does not favour.*

In the 1949 and 1960 editions of his book titled *Historical Geology*, Carl O. Dunbar correctly pointed out that "fossils provide the only historical, documentary evidence that life has evolved from simpler to more complex forms."[51] However, the third edition (1969) coauthored with Karl M. Waage omits the section "Documents of Evolution" that contained this information. If, according to some evolutionists, the fossil record in favour of evolutionism improved steadily over the years, it would be silly to have suppressed this pertinent information. The fossil record did not improve as claimed, and the few discoveries so far are disputable. This negates another of Feynman's principles of scientific integrity:

> If you've made up your mind to test a theory, or you want to explain some idea, you should always decide to publish it whichever way it comes out. If we only publish results of a certain kind, we can make the argument look good. We must publish *both* kinds of results.[52]

NASIM, for instance, in *Science, Evolution, and Creationism* discusses at length fossil findings such as Tiktaalik and Archaeopteryx that appear to favour evolutionism, but the organization does not mention findings that dinosaurs ate grass[53] or that their remains contained soft tissue, blood vessels, and cells.[54] These discoveries, which I consider in *The Death of Evolution*,[55] demand an explanation because they contradict the evolutionist worldview but are consistent with creationism. Scientific integrity demands that NASIM report results that challenge the Darwinian theory of evolution.

The late Stephen Jay Gould described this trick of hiding information unfavourable to evolutionism: "The extreme rarity of transitional forms in the fossil record persists as a trade secret of paleontology."[56] There are no secrets in science, but there are in pseudoscience or cargo-cult science!

♣ *Evolutionism disregards common sense and truth.*

Truth becomes secondary when data are chosen to fit or justify a preconceived conclusion. Under the spell of evolutionism, chaos instills organization, disorder promotes order, chance inspires design, matter invents information, non-intelligence creates intelligence, and the material paves way for the immaterial. To support these fantasies, some scholars select only certain data. Scott C. Todd thus insists that "Even if all data point to an intelligent designer, such an hypothesis is excluded from science because it is not naturalistic."[57] In other words, the ultimate goal of some modern scientists is not to arrive at the truth but simply to justify their philosophical preference. This is cargo-cult science. Remember Feynman's warning: "The first principle is that you must not fool yourself—and you are the easiest person to fool." To avoid such dishonesty and self-deception, Feynman admonishes:

> Details that could throw doubt on your interpretation must be given, if you know them. You must do the best you can—if you know anything at all wrong, or possibly wrong—to explain it. If you make a theory, for example, and advertise it, or put it out, then you must also put down all the facts that disagree with it, as well as those that agree with it. There is also a more subtle problem. When you have put a lot of ideas together to make an elaborate theory, you want to make sure, when explaining what it fits, that those things it fits are not just the things that gave you the idea for the theory: but that the finished theory makes something else come out right, in addition.[58]

Scientists are not philosophers; therefore, they should not engage in the philosophical interpretation of scientific data. "It is not," writes Richard Lewontin, "that the methods and institutions of science somehow compel us to accept a material explanation of the phenomenal world, but, on the contrary, that we are forced by our *a priori* adherence to material causes."[59] It is preferable, therefore, to have no scientific explanation at all about our origins than to accept the wrong one imposed upon us by Darwin and his disciples.

♣ *Evolutionism has the clout to silence the truth and anyone who dares to question.*

Richard Milton, a science journalist for *Mensa* magazine, explains how some leading evolutionists were upset with him for writing his book *Shattering the Myths of Darwinism*:

> When *Shattering the Myths of Darwinism* was published, I expected it to arouse controversy, because it reports on scientific research that is itself controversial and because it deals with Darwinism—always a touchy subject with the biology establishment. I didn't expect science to welcome an inquisitive reporter, but I did expect the controversy to be conducted at a rational level, that people would rightly demand to inspect my evidence more closely and question me on the correctness of this or that fact. To my horror, I found that instead of challenging me, orthodox scientists simply set about seeing me off "their" property.[60]

Milton goes on to describe how Richard Dawkins devoted two-thirds of his *New Statesman* review to attacking Milton's British publisher, Fourth Estate, for being irresponsible in accepting a book that criticized Darwinism. Dawkins lambasted the work as "'loony,' 'stupid,' 'drivel'" and its author as a "'harmless fruitcake' who 'needs psychiatric help.'" Below is a summary of Milton's concerns about such an ideologically driven attack:

> Dawkins is employed at one of Britain's most distinguished universities and is responsible for the education of future generations of students. Yet this is not the language of a responsible scientist and teacher. It is the language of a

religious fundamentalist whose faith has been profaned. *Nature* magazine, probably the most highly respected scientific magazine in the world, scented blood and joined in the frenzy. Its editor, John Maddox, ran a leading article that described me as believing science to be a myth (I don't), evolution to be false (I don't), and natural selection to be a pack of lies (I don't). It also magisterially rebuked the *Sunday Times* for daring to devote most of one of its main news pages to reporting the book's disclosures.[61]

"Fundamentalist" is a term that is supposed to characterize the religious and not the scientific world. This suggests that evolutionism operates more or less like a religious organization, with its own leaders and membership.

Apparently students are also intimidated by the disciples of evolutionism. Creationist Ariel A. Roth recounts the following about his graduate-school experience:

When I was a graduate student, the professor of evolution informed me that the faculty of the Department of Zoology was concerned about my creationistic beliefs. He wondered if I could explain them. I responded that I could see how a certain line of thought could lead to a belief in evolution, but that I had several questions about the theory. He was interested. One of the arguments I presented was that I could not understand how the turtle could have evolved from some other reptile without leaving fossil intermediates. The turtle is a unique organism, and in evolving this uniqueness—especially a shell—many intermediates would be involved, yet there is no such evidence in the fossil record. Paleontologists have found thousands of fossil turtles, some almost four meters long. They supposedly evolved some more than 200 million years ago, and in layers below where they first appear, we see no gradual sequence of the evolution of their peculiar shell. After discussing some other considerations, the professor seemed satisfied with my answers and agreed that evolutionary theory had some problems. Later I was told that the only reason the faculty allowed me to graduate was that they could not agree on what to do with me![62]

Scientific facts are transparent, but pseudoscience needs biased defense from its advocates. In Roth's case his evolutionist professor was finally led to admit another weakness of the Darwinian theory based on scientific evidence.

♣ *Evolutionism is a paradigm whose advocates are inconsistent.*

Evolutionism is a pseudoscience whose advocates are inconsistent in what they claim. For example, NASIM contends that "biologists are confident in their understanding of how evolution occurs."[63] Douglas J. Futuyma, however, reports that evolutionary theory, including the ideas of mutation, recombination, natural selection, genetic drift, and isolation, is subject to debate. He mentions two major arguments about evolutionary theory in scientific circles: (1) philosophical arguments about whether or not evolutionary theory qualifies as a scientific theory; and (2) substantive arguments about the details of the theory and their adequacy to explain observed phenomena. Futuyma also indicates that paleontologists such as Stephen Jay Gould have questioned neo-Darwinian theory's sufficiency in accounting for the broad panorama of historical evolution.[64] Kim Sterelny's *Dawkins vs. Gould: Survival of the Fittest* discusses the savage battle over the mechanism of evolution that continues to rage even after Gould's death in 2002.[65] These facts must be taken into serious consideration.

I would suggest that the bacteria-to-human model of evolution by natural selection does not qualify as a scientific theory. The explanation, calculation, and interpretation of evolutionary events are always based on speculative assumptions. This is the main reason why modern scientists should adhere, in the presentation of data pertaining to evolution, to Feynman's guidelines. Unfortunately, it appears that most evolutionists are prepared to give up the kind of scientific integrity that Feynman advocates in order to maintain their professional positions and financial support.

Philosophers Michael Ruse and Antony Flew offer important insights into the creationism-evolutionism debate. Both agree that evolutionism is a form of religion. In this regard Flew asserts:

> You might ask how I, a philosopher, could speak to issues treated by scientists. The best way to answer this is with another question. Are we engaging in science or philosophy here? When you study the interaction of two physical bodies, for instance, two subatomic particles, you are engaged in science. When you ask how it is that those subatomic particles—or anything physical—could exist and why, you are engaged in philosophy. When you draw philosophical conclusions from scientific data, then you are thinking as a philosopher.[66]

Flew and Ruse, however, have reached different conclusions. Previously an atheist, Flew concluded after fifty years that there is a God, insisting that his change of mind came about for purely scientific reasons: "I must stress that my discovery of the Divine has proceeded on a purely natural level, without any reference to supernatural phenomena."[67] Ruse, on the other hand, was a Christian before he became an atheist. Unlike Dawkins, Ruse has a genuinely liberal attitude toward the creationism-evolutionism controversy. He thus admits to finding Dawkins' *The God Delusion* embarrassing yet endorses evolutionism by asserting:

> The theory simply is not falsifiable. There is nothing conceivable which could exist and which could refute the theory. Central to Darwinism is the widespread existence of organic adaptation; but no matter how grotesque or abnormal, the Darwinians will always think up an adaptive explanatory analysis.[68]

Since there is no way to falsify their theory, evolutionists tell stories that fit their preconceived ideas and preferred conclusions. We are indebted to scientists for discoveries such as DNA, but scientists should not dictate whether DNA points to intelligent or unintelligent design. Scientists like Richard Dawkins who are responsible for the public understanding of science ought not use their positions to promote atheism.

In sum, every scientific study related to the origin of life and the universe has limitations because such origins are unknowable through science. Any scientific investigation pertaining to this subject proceeds on the basis of opinion or belief. With reference to the origin of life and, hence, of species, belief in bacteria-to-human evolution by natural selection (evolutionism) parallels belief in special creation (creationism).

Scientists are not divided over the field of evolution that biologists call microevolution. Medical and agricultural applications of evolutionary concepts are all confined to this framework, which has no religious implications. However, when the envisioned change is from bacteria to human, the issue of origin becomes important as a philosophical rather than scientific matter.

The field of evolution that biologists call macroevolution has transparently religious implications. Belief and interpretation based on the evolution model constitutes evolutionism, while that based on the creation paradigm is creationism. It must be stressed again that evolutionism is a secular religion whose proponents maintain a belief in chance and materialism. The evolutionists' insistence on delving into domains unsolvable as a scientific problem lays the foundation for pseudoscientific delusion.

NOTES

1. *Science, Evolution, and Creationism* (Washington, D.C.: National Academies Press, 2008), p. 12.
2. "The God Delusion?" 15 November 2006. *http://www.intentblog. com/archives/2006/11//the_god_delusio.html.* Retrieved 25 August 2008.
3. "E-Mail Evolution, *"The American Spectator,* July 1994, p. 16.
4. *Science, Evolution, and Creationism,* p. 7.
5. Ibid.
6. Ibid, p. 49.
7. Ibid, p. 7.
8. "29 + Evidences for Macroevolution: The Scientific Case for Common Descent." *http://www.talkorigins.org/faqs/comdesc/.* Retrieved 2 September 2008.
9. "Cover Story: Uprooting Darwin's Tree", *NewScientist,* 24-30 January 2009, p. 34.
10. Ibid.
11. The Origin-of-Life Prize ®, About the Gene Emergence Project. *http://www.us.net/life/rul_abou.htm.* Retrieved 14 March 2011.
12. The Origin-of-Life Prize ®, Description of the Prize. *http://www.us.net/life/rul_desc.htm.* Retrieved 14 March 2011.
13. Ibid.
14. The Origin-of-Life Prize ®, Submissions. *http://lifeorigin.org/rul_subm.htm* Retrieved 14 March 2011.
15. The Origin-of-Life Prize ®, News. *http://www.us.net/life/rul_news.htm.* Retrieved 20 August 2008.
16. *Science, Evolution, and Creationism,* p. 22.
17. See The Origin-of-Life Prize ®, Suggested Texts. *http://www.us.net/life/rul_sugg.htm.* Retrieved 14 March 2011; The Origin-of-Life Prize ®, Definitions. *http://www.us.net/life/rul_defi.htm.* Retrieved 14 March 2011; and The Origin-of-Life Prize ®, Discussion. *http://www.us.net/life/rul_disc.htm.* Retrieved 14 March 2011.
18. "Information Theory, Evolution, and the Origin of Life." *http://www.cynthiayockey.com/pages/1/index.htm.* Retrieved 20 August 2008.

19. "The Origin of the Universe Does Not Seem to Me to Be a Scientific Question," *Cosmos, Bios, Theos*, ed. Henry Margenau and Roy Abraham Varghese (La Salle, IL: Open Court, 1992), p. 166.

20. *The Origin of Species* (New York: Signet, 2003), pp. 39-40.

21. Ibid, pp. 46-47.

22. Ibid, p. 54.

23. Ibid, p. 57.

24. Ibid, pp. 452, 456, 458.

25. "Origin of Species," *NewScientist*, 14 May 1981, p. 452.

26. "Creation Is Supported by All the Data So Far," *Cosmos, Bios, Theos*, p. 78.

27. *Creation: Facts of Life* (Green Forest, AR: Master Books, 2006,), p. 101.

28. Introduction, *The Origin of Species* (London: J. M. Dent & Sons, 1971, pp. x-xi.

29. *Science, Evolution, and Creationism*, p. 49.

30. "The Dangers of Creationism in Education," Council of Europe Parliamentary Assembly, 2007. *http://assembly.coe.int/Documents/WorkingDocs/Doc07/EDOC11297.htm*. Retrieved on Retrieved 14 March 2011.

31. Quoted in John Allemang, "The Infinite Wisdom of Richard Dawkins," *The Globe and Mail*, 23 June 2007, Focus F$_3$.

32. *The Blind Watchmaker* (London: Penguin, 2006), p. xv.

33. *Creation: Facts of Life*, p. 11.

34. *Physics Today* 35 (October 1982), pp. 15, 103.

35. "The Dangers of Creationism in Education," A.4.

36. Ibid, A.14.

37. Ibid, A.5.

38. Union of Concerned Scientists, "Restoring Scientific Integrity in Policymaking." *http://www.ucsusa.org/scientific_integrity/interference/scientists-signon*-statement.html. Retrieved on 30 June 2008.

39. Wikipedia, Richard Feynman, *http://en.wikipedia.org/wiki/Richard_Feynman* Encyclopedia Britannica describes Feynman as the "American theoretical physicist who was widely regarded

as the most brilliant, influential, and iconoclastic figure in his field in the post-World War 11 era. Feynman was highly regarded by every scientific community in the world. He was a member of the American Physical Society; the American Association for the Advancement of Science, the National Academy of Science, and in 1965 was elected as a foreign member of the Royal Society, Great Britain.

40. Richard P. Feynman and Ralph Leighton, "Cargo[-]Cult Science," *Surely You're Joking, Mr. Feynman!: Adventures of a Curious Character* (New York: W. W. Norton, 1985), p. 341. The *Concise Oxford English Dictionary* (2004) defines "cargo cult" as "A system of beliefs based around the expected arrival of benevolent spirits in aircraft or ships bringing cargoes of food and other goods."

41. Ibid, p. 342.

42. Ibid, p. 340.

43. *Evidence for Truth: Science* (Guildford, Surrey: Eagle, 1998), pp. 148-49.

44. "The Evolutionary Vision," *Issues in Evolution,* vol. 3, ed. Sol Tax and Charles Callender (Chicago: University of Chicago Press, 1960), pp. 252-53.

45. *Tornado in a Junkyard* (Arlington, MA: Refuge Books, 1999), p. 233.

46. *Evolution: A Theory in Crisis* (Bethesda, MD: Adler & Adler, 1986), p. 16.

47. "Creationists Weaken Society's Trust in Scientists," *Nature*, 7 December 2006, p. 679.

48. *Science, Evolution, and Creationism*, pp. 40, 42.

49. Ibid, p. 43.

50. "On Methods of Evolutionary Biology and Anthropology," *American Scientist* 45 (1957), p. 388.

51. *Historical Geology* (New York: John Wiley, 1949), p. 52; *Historical Geology*, 2nd ed. (New York: John Wiley, 1960), p. 47.

52. "Cargo[-]Cult Science," p. 343.

53. Dolores R. Piperno and Hans-Dieter Sues, "Dinosaurs Dined on Grass," *Science,* 18 November 2005, p. 1126.

54. Mary H. Schweltzer, Jennifer L. Wittmeyer, John R. Horner, and Jan K. Toporski, "Soft-Tissue Vessels and Cellular Preservation in Tyrannosaurus Rex," *Science*, 25 March 2005, pp. 1952-55.

55. *The Death of Evolution: God's Creation Patent and Seal* (Longwood, FL: Xulon Press, 2007), pp. 36-40.

56. "This View of Life: Evolution's Erratic Pace," *Natural History*, 86.5 (May 1977), p. 14.

57. "A View from Kansas on That Evolution Debate," *Nature*, 30 September 1999, p. 423.

58. "Cargo[-]Cult Science," p. 341.

59. "Billions and Billions of Demons," *The New York Review*, 9 January 1997, p. 31.

60. *Shattering the Myths of Darwinism* (Rochester,Vermont: Park Street Press, 1997), p. ix.

61. Ibid, p. x.

62. *Origins: Linking Science and Scripture* (Hagerstown, MD: Herald Publishing, 1998), p. 182.

63. *Science, Evolution, and Creationism*, p. xiii.

64. *Science on Trial: The Case for Evolution* (New York: Pantheon, 1983), pp. 170-71, 173.

65. *Dawkins vs. Gould: Survival of the Fittest* (Cambridge: Icon Books, 2007).

66. *There Is a God: How the World's Most Notorious Atheist Changed His Mind* (New York: Harper, 2007), p. 89.

67. Ibid, p. 93.

68. "Darwin's Theory: An Exercise in Science," *NewScientist*, 25 June 1981, p. 828.

CHAPTER 2

DEBUNKING FALSEHOOD: THE EARTH'S AGE NOT IMPLIED IN THE BIBLE

The most conclusive argument against Genesis is its chronology: Genesis does not suggest an age of the Earth as old as 3,500 million years; indeed, the usual fundamentalist figure is about 6,000 years, although the creationists of Arkansas in 1981 preferred 'a relatively recent inception of the earth and living kinds'. That 'relatively recent' was chosen to contradict the scientific chronology, as does the figure of 6,000 years. The radioisotopic evidence of the antiquity of the earth, therefore, provides a decisive argument against Genesis.[1]

—Mark Ridley

The ideas of "creation science" derive from the conviction that God created the universe—including humans and other living things—all at once in the relatively recent past. However, scientists from many fields have examined these ideas and have found them to be scientifically insupportable. For example, evidence for a very young earth is incompatible with many different methods of establishing the age of rocks.[2]

—NASIM

THE MISINTERPRETATION OF BIBLICAL COSMOLOGY

Genesis posits neither an old (millions of years) nor a young (thousands of years) earth. The Scriptures say nothing about the age of the earth or universe, but scholars do. The ages cited in Mark Ridley's above remark depend on which group of scholars you ask. Unfortunately, creationists compound the problem because some profess a young earth and others an old earth. This difference in interpretation prompts evolutionists to dismiss the Genesis account as an ancient myth and instead credit the evolutionary worldview as a scientific fact.

First, we must clarify the meaning of "age." The *Oxford English Reference Dictionary* defines the word as signifying "the length of time that a person or thing has existed or is likely to exist."[3] Let us take a building as an example. A real-estate agent—I'll call him Sam—who is also a certified geologist conceivably might specify two ages: the structural age and the matrix age. In his role as a real-estate agent, depending on when the building was completed, Sam might cite an age of, say, 50 years. This is the structural age. In his other role as a geologist Sam might take a rock sample from the building, subject it to radiometric dating, and cite an age of millions of years. This is the matrix age. Therefore, for the same building you have two different ages depending upon your interest. If you are looking at it from a real-estate point of view, the building is young, but from a geological point of view it is old. Which is correct? The question of the earth/universe's age can be addressed from both points of view.

A further example, consider the formation of a canyon due to volcanic activities, such as that formed at the eruption of Mount St. Helens in 1980. On this event, James Perloff in his book *Tornado in a Junkyard* writes:

> In 1980, the eruption of Mount St. Helens created strata up to 600 feet high; within five years they had hardened into rock. The volcano toppled 150 square miles of forest in six minutes. Its mud flows eroded a canyon system as deep as 140 feet; later on, a small creek was found

running through it. Had this eruption occurred a couple of centuries ago, a modern uniformitarian geologist, inspecting these features, might conclude that the strata had formed over eons, and that the creek had carved the canyon.[4]

Within five years the strata had hardened into rock, hence, its structural age is around 5 years. Radiometric dating of the lava dome of Mount St. Helens gave a value of 2.8 million years[10]; this is, in the present context, its matrix age. For the same formation, we have two ages: a structural age of about 5 years and a matrix age of about 3 million years. Both ages are describing the same thing from different perspectives: one looks at the creation of the canyon and the other looks at the constituents of the canyon. These are the age categories of concern to the creationists and evolutionists. Both are different facts of the same object!

The six days mentioned in biblical cosmology (Gen. 1; Exod. 20:8-11) relate to neither the structural nor the matrix age of the earth. They describe instead God's span of time for preparing the earth to produce and accommodate the basic kinds of living organisms, including plants, animals, and humankind.

Does the Bible actually claim that the earth is 6,000 years old? The truth is that God is silent about the earth's age. The Scriptures indicate the life spans of people but not the age of earth or the cosmos. Nowhere in Genesis 1 is there any indication of the earth's age. The earth's age is irrelevant to the creationist worldview but is crucial for the evolutionist worldview that presupposes eons.

In my book *The Death of Evolution* I identify two categories of time. The first is uncaused time ($t < 0$) before the world began, which is consistent with God's attribute of existing from time immemorial or all eternity.[5] The other, caused time, corresponds to $t = 0$, which marks the beginning of the physical world. From a theological point of view, the beginning of physical time $t = 0$ marks the beginning of the earth's structural age. From the scientific point of view, $t = 0$ marks the Big Bang when the universe exploded from dense matter. If this dense matter existed as theorized, then it is part of God's creation since God is both the omnific Beginning and End.

On the first day of physical time, four dimensions—the heavens, the formless earth in a matrix of water, darkness, and light—originating from "nothing" became "something" in our physical universe. In reference to the geosphere, the structural age is zero. This postulate is consistent with Yale University physicist Henry Margenau's assertion that "God created the universe out of nothing in an act which also brought time into existence."[6] I now will explain why Margenau's statement makes sense.

In reference to light, for example, God is the embodiment of light (Isa. 60:19-20; James 1:17). There is credible historical evidence for this assertion. When Moses came down from Mount Sinai with the Ten Commandments after witnessing the presence of God, his face beamed with radiance. It was so bright to human eyes that he had to wear a veil over his face when engaged in public duty (Exod. 34:29-35). Since light is one of the attributes of God (John 1:5), it existed before the beginning of physical time. In preparing the earth, sea, and sky for habitation, God's command "Let there be light" inaugurated light, one of the essential components in forming a physically life-sustaining environment. This command, "Let there be light," defines the beginning of time and assigns an age of zero to light, although it preexisted. By way of analogy, how could one date the age of light that electricity generates?

The elements enumerated in Genesis 1:1 appeared in the physical world (Heb. 11:3) as raw materials (matter and energy) for earth's construction to be accomplished in six days. They are like the architect's blueprints for an edifice. In order for a foundation to be laid, the raw materials must already be in place. This principle is consistent with scriptural assertion: "In the beginning you laid the foundations of the earth, and the heavens are the works of your hand" (Ps. 102:25, NIV). Therefore, Genesis is a narrative of God's provision of elements for sustaining life on earth. If anything, the account only gives only a rough and unspecified idea of when the planetary environment became functional.

Radiometric techniques for determining the earth's age easily yield wrong results because no one knows the planet's primordial conditions. Ex-evolutionist Gary Parker thus comments:

> One of the tensest moments for me came when we started discussing uranium-lead and other radiometric methods for estimating the age of the Earth. I just knew all the creationists' arguments would be shot down and crumbled, but just the opposite happened. In one graduate class, the professor told us we didn't have to memorize the dates of the geologic systems, since they were far too uncertain and conflicting. Then in geophysics we went over all the assumptions that go into radiometric dating. Afterwards, the professor said something like this, "If a fundamentalist ever got hold of this stuff, he would make havoc out of the radiometric dating system. So, keep the faith." That's what he told us, "Keep the faith." If it was a matter of keeping faith, I now had another faith I preferred to keep.[7]

This trade secret of empirical uncertainty presumably explains why evolutionists cite an age for the universe without bothering to provide the magnitude of error associated with their estimates.

Recently Michael W. Robbins published an article titled "How We Know Earth's Age" in which geologist Paul Renne, Director of the Berkeley Geochronology Center, is cited as remarking: "Argon-argon dating is powerful and widely applicable. It is how scientists know that Earth is 4.5 billion years old and not 6,004 years, as some biblical literalists believe, or 5,765 years old, as some Orthodox rabbis believe."[8] When a recognized scientist makes such a claim, one presumes zero uncertainty. However, the preceding quotation is a clear example of selective sampling.

People who believe everything that scientists claim as facts are easily misled, but those who question suspicious claims discover the truth. For example, when both Richard Dawkins and *Nature* magazine told Richard Milton, author of *Shattering the Myths of Darwinism*, that he believed in the estimate that the earth is merely a few thousand years old, Milton retorted:

> I do not believe that the Earth is only a few thousand years old. I present evidence that currently accepted methods of dating are seriously flawed and are supported by Darwinists only because they provide the billions of years required by Darwinist theories. Because radioactive dating methods are scientifically unreliable, it is at present impossible to say with any confidence how old the Earth is.[9]

Here Milton adheres to the silence of God—i.e., that no human being knows the correct age of the earth. In this connection Peter Grace points out the following:

> Evolutionist and creationist speculations necessarily assume certain factors to be constant, e.g. the rate of decay of a radioisotope, the amount of a substance already present at the onset of any given chemical or organic process, the amount of pollution that has taken place in the course of the process. And because these variables have happened in the past and therefore are unobserved, the calculation remains speculative.[10]

Scientists are, therefore, just as ignorant as theologians of the correct age of the earth.

Bruce A. Malone has illustrated how inaccurate the age estimates can be by using the following formulas:

$$\text{TIME} = \text{AMOUNT} \div \text{RATE}$$

$$\text{AMOUNT} = \text{MEASURED—INITIAL—CONTAMINATION}[11]$$

In all dating methods the present amount and present rate are the only things that scientists know. The initial amount, estimate of contamination, and average rate are all unprovable assumptions about the unknown past. The age determination is thus flawed by the wrong set of assumptions. Let us examine how this can influence age calculations by using Malone's approach.

- Suppose John's home is measured to be 15 km away, and he approaches you at 6:00 a.m. walking at a constant speed of 3 km/hr. Based on the formula above, we calculate that John left home 5 hours earlier. You have determined the time (age) of his trip. You assumed that John started from home.

- Suppose John was instead coming from a nearby coffee shop less than 1 km away. Then the initial amount in your age calculation is wrong because you have the wrong starting point. You thus end up with the wrong time (age) of his trip.

- Suppose John started from home but took a shortcut instead of following the normal route of 15 km. In this case there is contamination of the total amount. Again, assuming that he maintained the same speed of 3 km/hr, your time (age) estimate is wrong, although the formula is correct.

- Suppose John started from home but jogged most of his way before walking toward you. Your time (age) measurement for his trip is wrong since the speed (present rate) of 3 km/hr was inconsistent, although you used the correct formula.

Use of different methods to determine the earth's age will require different sets of assumptions. Consistency or precision in the values obtained does not guarantee accuracy in the age deduced. Nobody knows enough about events at the beginning of time to draw any valid conclusion. Evolutionists use assumptions that justify an old earth, while some creationists use assumptions that justify a relatively young earth. Another important point is that, depending on the physical phenomenon of interest, the earth may appear either young or old. The essential truth is that we do not need to know the age of the earth for science to be what it is today. Therefore, creationists and evolutionists should stop quibbling about the earth's age.

In sum, the earth has both a structural and a matrix age. The age of approximately 4.5 billion years that evolutionists cite is an estimate of the matrix age. This estimate is based on extrapolations and assumptions about the unknown past that cannot be validated; therefore, its accuracy is limited by human imagination. The age of 6,000-10,000 years that some creationists deduce from the Scriptures is an estimate of the earth's structural age.[12] The Bible is silent about both ages, though on many occasions it gives the life spans of prominent people because it deems such information of historical importance.

I have some reservations about Kenneth R. Miller's view that a truly scientific attempt to answer the question of biological origins would begin by examining the age of the earth and reviewing the scientific techniques used by geologists to determine the ages of rocks and fossils. My reasons are: (a) the geological ages are unreliable because they are based on assumptions that cannot be verified; (b) the indicated ages of rocks and fossils are based on circular reasoning; and (c) a knowledge of the earth's age does not prove either creation or evolution. Miller writes, "Ignoring the age of the earth while attempting to teach students natural history makes about as much sense as trying to teach American history without telling students that the American Revolution began in 1775, which is to say, no sense at all."[13] This is an invalid analogy, however, because there were no human beings when the earth began. Besides, science is primarily concerned with how nature works or functions. Accordingly, the Scriptures talk about how God commissioned humankind as a custodian to explore the earth. The age of the earth is not the determining factor of whether we evolved by chance or were created in the image of God.

As John F. Haught points out, "Creationism turns a font of holy wisdom into a mundane treatise to be placed in competition with shallow scientific attempts to explain things."[14] Creationists should stick to direct claims in the Scriptures and refrain from reaching false conclusions based on deductions. The Scriptures are rich with direct claims:

> For this is what the LORD says—He who created the
> heavens, He is God; He who fashioned and made the
> earth, He founded it; He did not create it to be empty,
> but formed it to be inhabited—He says: "I am the LORD,
> there is no other." (Isa. 45:18 NIV)

> It is I who made the earth and created mankind upon it.
> My own hands stretched out the heavens: I marshalled
> their starry hosts. (Isa. 45:12 NIV)

The Scriptures maintain that the earth was designed purposely to be inhabited. The inhabitants and the environment were made in six days (Gen.1:1-31). God personally made this claim before a gathering of ancient Israelites and wrote it down as a commandment (Exod. 20: 8-11; 31:12-18). A commandment is a statement of fact, not myth. And as 1969 Nobel laureate chemist D. H. R. Barton asserts, "The ultimate truth is God."[15] This is an historical event that generations cannot erase. There are those who do not believe that the Holocaust occurred, despite the testimonies of people who survived it as living witnesses. Whether sceptics believe it or not, the truth remains.

I conclude this chapter with a tribute to distinguished scientist Sir John Ambrose Fleming, remembered for his invention of the thermionic wireless valve and diode. Sir Fleming was a devout Christian who, with Douglas Dewar and Bernard Acworth, established the Evolution Protest Movement in 1932 and was its first president.[16] He was also president of the Victoria Institute and the Philosophical Society of Great Britain, a scientific community that expressed its confidence in the Holy Scriptures. Their manifesto, issued in 1864 and titled "The Declaration of Students of the Natural and Physical Sciences," was signed by 717 scientists, including 86 Fellows of the Royal Society. It reads:

> We, the undersigned Students of the Natural Sciences,
> desire to express our sincere regret that researches into
> scientific truth are perverted by some in our own times
> into an occasion for casting doubt upon the Truth and

Authenticity of the Holy Scriptures . . . . We believe that it is the duty of every Scientific Student to investigate nature simply for the purpose of elucidating truth, and that if he finds that some of his results appear to be in contradiction to the Written Word, or rather to his own *interpretations* of it, which may be erroneous, he should not presumptuously affirm that his own conclusions must be right, and the statements of Scripture wrong; rather, leave the two side by side till it shall please God to allow us to see the manner in which they may be reconciled; and, instead of insisting upon the seeming differences between Science and the Scriptures, it would be as well to rest in faith upon the points in which they agree.[17]

These words of wisdom from accomplished scientists of the past apply to the modern world. The apparent conflict between science and the Scriptures is the result of subtle misunderstanding of the latter by both theologians and scientists. The Holy Scriptures do not give us the age of the earth. Science can neither establish nor disprove the structural age of the earth; hence, the only logical conclusion is that the earth is neither young nor old.

NOTES

1. *The Problems of Evolution* (Oxford: Oxford University Press, 1985), p. 2.
2. "Teaching About Evolution and the Nature of Science (1998)." http://www.nap.edu/openbook.php?isbn=0309063647&page=55. Retrieved on 3 April 2011.
3. *The Oxford English Reference Dictionary*, 2nd ed. (Oxford: Oxford University Press, 2001), p. 24.
4. *Tornado in a Junkyard* (Arlington, MA: Refuge Books, 1999), p. 159.
5. *The Death of Evolution: God's Creation Patent and Seal* (Longwood, FL: Xulon Press, 2007), pp. 175-78.
6. "The Laws of Nature Are Created by God," *Cosmos, Bios, Theos*, ed. Henry Margenau and Roy Abraham Varghese (La Salle, IL: Open Court, 1992), p. 57.
7. *From Evolution to Creation: A Personal Testimony* (Florence, KY: Answers in Genesis, 2000), p. 10.
8. "How We Know Earth's Age," *Discover*, March 2006, p. 22.
9. *Shattering the Myths of Darwinism* (Rochester, Vermont: Park Street Press, 1997), p. ix.
10. "The Death of Evolution?" http://www.theotokos.org.uk/pages/creation/pgrace/deathevo.html. Retrieved on 3 April 2011.
11. *Search for the Truth* (Midland, MI: Search for the Truth Ministries, 2001), pp. iv-9.
12. See E. C. Ashby, *Understanding the Creation-Evolution Controversy* (Ozark, AL: ACW Press, 2005), pp. 27-28.
13. "Of Pandas and People: A Brief Critique." http://www.kcfs.org/pandas.html. Retrieved on 3 April 2011.
14. *Science and Religion* (New York: Paulist Press, 1995), p. 53.
15. "The Ultimate Truth Is God," *Cosmos, Bios, Theos*, p.166.
16. Henry M. Morris, *Men of Science, Men of God* (Green Forest, State?: Master Books, 1988), p. 75.
17. "Scientia Scientiarum," *Journal of the Transactions of the Victoria Institute* (1865), http://www.creationism.org/victoria/VictoriaInst1866_pg005.htm. Retrieved on 3 April 2011.

CHAPTER 3

RESOLVING THE MICRO-MACRO EVOLUTION DILEMMA

In the twenties, the Russian entomologist Iurii Filipchenko divided evolution up into two basic categories: *microevolution* and *macroevolution*. The terms, borrowed from the Greek words for 'small' and 'large,' simply distinguished small-scale, incremental evolution (like a mutation that changes the colour of a pupil)—microevolution, from the bigger, more dramatic changes of macroevolution, as when one species transform into a new one.[1]

—Danny Vendramini

Macroevolution: A vague term generally used to refer to evolution on a grand scale, or over long periods of time. There is no precise scientific definition for this term, but it is often used to refer to the emergence or modification of taxa at or above the genus level. The origin or adaptive radiation of a higher taxon, such as vertebrates, could be called a macroevolutionary event.[2]

—Glossary Evolution Library

ETYMOLOGY AND INCONSISTENCIES

In the current creationism-evolutionism controversy, creationists accept only microevolution or bacteria-to-bacteria evolution as a scientific fact, but evolutionists insist that macroevolution or bacteria-to-human evolution is also a scientific fact. The view that it requires millions of years to observe morphological transitions between organisms (macroevolution) is based on blind faith and thus is completely unscientific. A review of relevant terms in evolutionary theory is necessary to distinguish between proper science and scientific dogma or pseudoscience.

Leslie Alan Horvitz provides the following concise history of the terms microevolution and macroevolution.

> The terms *macroevolution* and *microevolution* were first coined in 1927 by Russian entomologist (insect researcher) Iurii Filipchenko, who was trying to reconcile evolution and Mendalian genetics. (He was not, however, a Darwinian.) His student, geneticist and zoologist Theodosius Dobzhansky, introduced the terms to the United States in his 1937 book *Genetics and the Origin of Species,* which he began by saying that "we are compelled at the present level of knowledge reluctantly to put a sign of equality between the mechanisms of macro—and microevolution." His reluctance stems from his realization that he was proposing a theory at odds with that of his beloved teacher.[3]

Today scholars present evolution as consisting of the two autonomous fields, microevolution and macroevolution. Macroevolution is a multidisciplinary field that relies on contributions from biologists, geneticists, paleontologists, molecular scientists, geologists, and astrophysicists. From the biological standpoint,

Biological Evolution = Microevolution + Macroevolution

According to E. C. Ashby, one of the most important problems that has caused division between creationists and evolutionists has been the inadequate definition of terms whenever the subject of creation/evolution is discussed.[4] Evolutionists are not unanimous in their explanation of microevolution and macroevolution. For instance, Austin Cline contends that "Macroevolution is merely the result of a lot of microevolution over a long period of time."[5] However, Kim Sterelny asserts that "macroevolution is not just microevolution scaled up."[6] Biologists, therefore, are inconsistent. The evolutionary paradigm is plagued by contradictions of this sort. In fact, because of evolutionary biology the very definition of science is in question. Courts now have to rule on what distinguishes scientific fact from scientific theory. Without a doubt the theory of evolution has raised questions about the methodological integrity of modern science.

Usage of the word "evolution" is also misleading. A. E. Wilder-Smith explains:

- The word "evolution" is often unconsciously employed in two different senses. First, "evolution" refers to the small, often genetically hidden variations present within every species and which may be discovered in the course of breeding. Or variations may be caused by mutations and selection, which may be inherited (microevolution).

- Second, the same term is applied to the transformation of one species into another higher species due allegedly to the accumulation of mutations (macroevolution or transformism) . . . . Many biologists attempt to confuse the issue by using the factuality of the above type of microevolution as a basis for proving the reality of macroevolution or transformism. For this reason they confuse and intermingle these two quite distinct evolutionary concepts and attempt to blend them into one concept. They use the one term (evolution) for the two concepts and then heap derision on all who deny "evolution" by pointing out that "evolution" (by which they mean microevolution) is a fact.[7]

Microevolution is thus a scientifically demonstrable fact, but macroevolution poses quite a different problem. In the following discussion I shall address the subject of macroevolution in light of Richard Feynman's vision of science in which scientific integrity overrides philosophical preference.

DEBUNKING MACROEVOLUTION

Feynman insists on an "extra type of integrity . . . when [one is] acting as a scientist." I shall also endorse Feynman's principle that, if a theory is to be tested or explained, scientists must publish *all* results of their inquiries.[8] What evolutionists unfortunately have done so far is to process information about mutation that leads to predetermined conclusions. In this regard evolutionist Pierre-P. Grassé writes:

> Biochemists and biologists who adhere blindly to the Darwinism theory search for results that will be in agreement with their theories and consequently orient their research in a given direction, whether it be in the field of ecology, ethology, sociology, demography (dynamics of populations), genetics (so-called evolutionary genetics), or paleontology. This intrusion of theories has unfortunate results: it deprives observations and experiments of their objectivity, makes them biased, and, moreover, creates false problems.[9]

Scientists can manipulate data to fit their theories and justify their philosophical preferences. For instance, without any hard empirical evidence about specific mechanisms, evolutionists still conclude that Darwinian bacteria-to-human evolution has occurred.[10] They brush aside contradictory evidence, since they believe that no credible and alternative theory exists. The irony is that essentially the same data evolutionists use to justify the evolutionary worldview also justify the creationist worldview.

Species were either created or have evolved. If scientists conclude in advance that species evolved and then seek data that justify their conclusion, they compromise the conventional standard of professional integrity. Science should follow the evidence faithfully regardless of whether people like its conclusions. Molecular biologist and physician Michael Denton states that the data on origins deserve both evolutionist and creationist interpretations.[11]

Microevolution and macroevolution, according to Ernst Mayr, are two autonomous fields.[12] Microevolution explains how life developed since its inception; macroevolution explains how life forms originated. Changes in the appearance of species and knowledge of their origin are two different issues. Problems arise when the two concepts are subsumed under a unifying theory of evolution.

The claim that microevolution leads to macroevolution is rejected by creationists and some evolutionists. David N. Menton thus writes:

> The very name "micro evolution" is intended to imply that it is this kind of variation that accumulates to produce macroevolution, though a growing number of evolutionists admit there is no evidence for this. Thus an observable phenomenon is extrapolated into an unobservable phenomenon for which there is no evidence, and then the latter is declared to be a "fact" on the strength of the former. It is this kind of limitless extrapolation that comprises much of the argument for evolution.[13]

Evolutionists use extrapolations to bridge the gap between science (facts) and pseudoscience (myths/beliefs/fantasies).

Extrapolations are useful in science when the mechanisms involved are well known. Darwin's evolutionary tree, however, includes numerous unknown branches, and his model thereby inhibits routine extrapolations. Percival Davis and Dean H. Kenyon explain the dilemma:

- People sometimes give the impression that any change is evidence for Darwinism. But Darwinism is not just any change. It is a very special kind—the transformation of one type of organism into another. Picture in your mind an evolutionary tree. The change produced by breeders is horizontal change, the flowering and elaboration of a single branch on that tree. What is needed, however, is vertical change leading up the evolutionary tree and creating a new branch.

- To put it another way, breeders can produce sweeter corn or fatter cattle, but they have not turned corn into another kind of plant or cattle into another kind of animal. What breeders accomplish is diversification within a given type, which occurs in microevolution. What is needed is the origin of new types, or *macroevolution*. Neo-Darwinism assumes that microevolution leads to macroevolution. To put that into English, it assumes that small-scale changes will gradually accumulate and produce large-scale changes.[14]

Extrapolations, like predeterminations, may lead us to faulty conclusions. Scientists are aware of these limitations, and in fact Darwin struggled with this point as Denton remarks:

> There was also the disturbing point, which Darwin was well aware of and had tried rather unconvincingly to dismiss at the end of Chapter Two of the *Origin*, that while breeding experiments and the domestication of animals had revealed that many species were capable of a considerable degree of change, they also revealed distinct limits in nearly every case beyond which no further change could ever be produced. Here then was a very well established fact, known for centuries, which seemed to run counter to his whole case, threatening not only his special theory—that one species could evolve into another—but also the plausibility of the extrapolation from micro to macroevolution, which, as we have seen, was largely based on an appeal to the remarkable degree of change achieved by artificial selection in a relatively short time.[15]

Microevolution and macroevolution are thus unconnected events. Microevolution is a scientifically demonstrable fact, but macroevolution poses quite a different problem. Darwin's theory suggests that all evolution is the result of random mutation and natural selection. In order to understand the microevolution-macroevolution controversy, we need some background knowledge about mutations.

From breeding and research studies on organisms such as the fruit fly, scientists have obtained empirical data about the nature and consequences of mutation. They have established the following facts:

1. *The influence and causes of mutation*:
   - Mutation is the central mechanism of evolution.[16]
   - Gene mutations occur when individual genes are altered or damaged from exposure to heat, chemicals, or radiation. Chromosome mutations occur when sections of DNA are duplicated, inverted, lost, or moved to another place in the DNA molecule.[17]
   - Mutants, whether they result in small, visible changes or large ones, are the material of evolution.[18]

2. *The dynamics and consequences of mutation*:
   - Mutations are random, not directed and rare.[19]
   - The great majority of mutations, certainly over 99 per cent, are harmful.[20]
   - Mutations cannot be controlled, and so natural selection must simply take what comes.[21]
   - Good mutations are extremely rare.[22]

Theodosius Dobzhansky asserts that "the mutation process alone, not corrected and guided by natural selection, would result in degeneration and extinction rather than improved adaptedness."[23] However, according to evolutionist Mark Ridley, "Natural selection takes time. For a mutant to increase in frequency from its initial rare state to become the normal gene in the whole population may take a few thousand generations."[24] Given this length of time, and the fact that mutations affect and are affected by many genes, Henry M. Morris argues:

> It would seem obvious that if any one mutation is highly likely to be deleterious, then since a changed characteristic requires the combined effects of many genes, and therefore many concurrent mutations, the probability of harmful effects is multiplied manyfold. Conversely, the probability of simultaneous good mutations (which are very, very rare) in all the genes which control a given character is reduced to practically zero.[25]

Under these circumstances evolution from simple to complex organisms is most unlikely, as has been demonstrated through laboratory studies on fruit flies and bacteria.

Based on empirical studies of fruit flies, the authors of *Of Pandas and People* conclude:

> There is no evidence mutations create new structures. They merely alter existing ones. Mutations have produced, for example, crumpled, oversized, and undersized wings. They have produced double sets of wings. But they have not created a new kind of wing. Nor have they transformed the fruit fly into a new kind of insect. Experiments have simply produced variations of fruit flies.[26]

They continue:

> Recall that the DNA is a molecular message. A mutation is a random change in the message, akin to a typing error. Typing errors rarely improve the quality of a written message; if too many occur, they may even destroy the information contained in it. Likewise, mutations rarely improve the quality of the DNA message, and too many may even be lethal to the organism.[27]

Scientists have been able to establish these facts because of the short life span of fruit flies, which enables scientists to observe many generations. In *Creation and Change,* Douglas F. Kelly discusses mutation studies undertaken by French zoologist Pierre-P. Grassé on bacteria with much shorter life spans than those of fruit flies. He writes:

> One bacteria generation lasts approximately 30 minutes. Hence they multiply 400,000 times faster than human generations. Researchers, therefore, can trace mutational change in bacteria in relatively brief compass equivalent to 3,500,000 years of change within the human species. But Grassé has found that these bacteria have not essentially changed during all these generations. In view of this empirical fact, is it reasonable to maintain that humankind has evolved during the same equivalent time period in which bacteria have been stable?[28]

These results show that even over eons new organisms have not developed as believed by Darwin and his disciples. Accordingly, Kelly concludes:

> Thus, hard genetic facts militate against evolution being possible through either recombination or mutations. The concept of an upward evolutionary scale of life is not grounded in empirical science[;] it is actually contrary to it. This means that the theory of evolution is really philosophy, not operational science.[29]

How do we explain these events in light of microevolution and macroevolution? Here is one response:

> All changes observed in the laboratory and the breeding pen are limited. They represent microevolution, not macroevolution. These limited changes do not accumulate the way Darwinian evolutionary theory requires in order to produce macro changes. The process that produces macroevolutionary changes must be different from any that geneticists have studied so far.[30]

This conclusion is consistent with evolutionist Sterelny's previously cited comment that macroevolution is not just scaled-up microevolution.

In contrast to the above conclusions, Francisco J. Ayala claims in *Darwin and Intelligent Design* that "natural selection is a creative process, although it does not create the raw materials upon which it acts."[31] To justify his claim, Ayala discusses how natural selection has produced bacterial cells resistant to streptomycin and not requiring histidine for growth. In these steps the organisms remained bacteria, resulting in bacteria-to-bacteria evolution or microevolution. Unless novelty implies antibiotic resistance, there is nothing of substance in his discussion to show how this relates to bacteria-to-human evolution or macroevolution. As the bacterial progeny are variants of their progenitors, they have not advanced an angstrom up the evolutionary tree that adorns biology textbooks.

Qualifying Ayala's perspective, ex-evolutionist Gary Parker makes the following sweeping remarks:

> Mutations are NOT genetic" script writers"; they are merely "typographic errors" in a genetic script that has already been written. Typically, a mutation changes only one letter in a genetic sentence averaging 1,500 letters long. To make evolution happen—*or even to make evolution a theory fit for scientific discussion*—evolutionists desperately need some kind of "genetic script writer" to *increase the quantity and quality of genetic* **INFORMATION.** Mutations have no ability to compose genetic sentences, no ability to produce genetic information, and hence no ability to make evolution happen at all.[32] (Emphasis his.)

If the raw material that evolves from a simple to complex state comes with mistakes in it and does not have the resources to fix it, what can? Natural selection? But natural selection takes time to act and only works on the faulty raw material as a selector in order to enable it to adapt. Natural selection is not an innovator. Richard C. Lewontin, a leading evolutionist from Harvard University concurs:

[N]atural selection operates essentially to enable the
organisms to maintain their state of adaptation rather than
to improve it . . . . [N]atural selection over the long run
does not seem to improve a species' chance of survival
but simply enables it to "track," or keep up with, the
constantly changing environment.[33]

The adaptation process, as Parker explains, results in minor
changes within the species boundary:

Natural selection does not lead to continual improvement
(evolution); it only helps to maintain features that organisms
already have (creation). Natural selection works only
because each kind was created with adaptation (design
features) and sufficient variety to multiply and fill the
earth in all its ecological and geographic variety.[34]

The above views on natural selection suggest that mutation
(the mechanism for evolution) followed by natural selection
(the designing instrumentality) is not the appropriate model for
macroevolution. Theodosius Dobzhansky ponders this matter in
Genetics and the Origin of Species:

A majority of mutations, both those arising in laboratories
and those stored in natural populations, produce
deteriorations [in] viability, hereditary diseases, and
monstrosities. Such changes, it would seem, can hardly
serve as evolutionary building blocks.[35]

If mutations are rare and harmful, how can they be the primary
source of our existence? If natural selection cannot intervene in case
studies on fruit flies and bacteria, why should it facilitate the quantum
leap from bacteria to human beings? Is this not pure delusion?

We cannot change empirical evidence, but we can change theory.
And a good place to start is our definition of the terms "microevolution"
and "macroevolution." In light of the empirical evidence, I shall now
propose some more useful and precise equivalents.

Darwin's tree of life posits that the bacteria-to-human evolution is directed upwards. We know from empirical science that mutation is random and not acted upon immediately by natural selection; hence nothing prevents the tree of life from growing downwards. Ex-evolutionist Gary Parker asserts:

> *Upward or downward?* Even more serious is the fact that mutations are "going the wrong way" as far as evolution is concerned. Almost every mutation we know is identified by the disease or abnormality that it causes. Creationists use *mutations* to explain the *origin of parasites and disease,* the *origin of hereditary defects,* and the *loss of traits.* In other words, time, chance, and random changes do just what we normally expect: tear things down and make matters worse. Using mutations to explain the *breakdown* of existing genetic order (creation-corruption) is quite the opposite of using mutations to explain the *build-up* of genetic order (evolution). Clearly, creation-corruption is the most direct inference from the effects of mutations that scientists actually observe.[36]

Consequently, if one adheres to Richard Feynman's rules for maintaining scientific integrity, evolution must be treated as bidirectional.

The bidirectional nature of evolution has been demonstrated in crossbreeding experiments. Harold G. Coffin, discussing Ernst Mayr's experiment with fruit flies, reports:

> A strain of the common fruit fly (*Drosophila melanogaster*) averaged 36 bristles on its body. Crossbreeding succeeded in lowering that number to 25, but then the experimentation ended because the line became sterile. In the other direction, success stopped at an increase of about 20 bristles before sterility halted the research.[37]

Concerning the same experiment Francis Hitching noted that "Then Mayr brought back non-selective breeding, letting nature take its course. Within five years, the bristle count was almost back to average."[38]

The above empirical results confirm that, for all intents and purposes, evolution must be treated as bidirectional. Accordingly, the proper representation of the preliminary dynamics in Darwinian evolution can be modeled conveniently by the Cartesian Coordinate System in which the positive y-axis represents gain in information and the negative y-axis represents loss of information. The positive x-axis or the line $y = 0$ represents neither loss nor gain in information and hypothetically represents the scenario at the organism's boundary. As a random process mutation can proceed upwards (positive y-axis) or downwards (negative y-axis). Since mutations are copying mistakes (loss of information) and beneficial ones are rare, the bulk of the raw material for natural selection proceeds along the negative y-axis. Mutations differ in their magnitudes. The micro ones survive, but the macro ones die out quickly.

For neo-Darwinian evolution to occur, we need to gain rather than lose information. In the domain of microevolution, when plants bearing fruits with seed are genetically manipulated, we end up with the same kind of plant bearing fruit but with no seeds. There is consequently some loss of information in the process because the information to produce seeds is no longer there. In addressing a similar point, James Perloff writes:

> Plant and animal breeders don't manufacture new genes; they simply use genetic information that is already there. In fact, their practices result in lost information. When you isolate, say, bulldogs, allowing them to mate only with other bulldogs, they stay bulldogs. They have lost the capacity to look any other way. Thus the creation of special varieties, such as Darwin's finches, means genetic information is gone—not gained, as evolution claims.[39]

How do we know from a molecular point of view whether a given event represents an information loss or gain? Lee Spetner asserts: "All point mutations that have been studied on the molecular level turn out to reduce the genetic information and not to increase it."[40] Here are some highlights of Spetner's investigation:

- If a mutation in the bacterium should happen to change the ribosome site where the streptomycin attaches, the drug will no longer have a place to which it can attach . . . . We see then that the mutation reduces the specificity of the ribosome protein, and that means *losing* genetic information. This loss of information leads to a *loss of sensitivity* to the drug and hence to resistance. Since the information loss is in the gene, the effect is heritable, and a whole strain of resistant bacteria can arise from the mutation.[41]

- Although such a mutation can have selective value, it decreases rather than increases the genetic information. It therefore cannot be typical of mutations that are supposed to help form small steps that make up macroevolution. Those steps must, on the average, add information. Even though resistance is gained, it's gained not by adding something, but by losing it. Rather than say that the bacterium gained resistance to the antibiotic, we would be more correct to say that it lost its sensitivity to it. It lost information . . . . Information cannot be built up by mutations that lose it. A business can't make money by losing it a little at a time.[42]

- How does an insect become resistant? It becomes resistant by losing its sensitivity to DDT. This loss is the result of a mutation that changes the site on the nerve cell at which the DDT molecule binds, preventing the DDT from binding. Any mutation that spoils the match between the DDT and the nerve cell will make the insect resistant. As with bacteria, resistance can come by reducing the specificity of the protein of the nerve cell.[43]

- Although there are circumstances where point mutations are good for the organism, all known point mutations lose information. Some microevolution may indeed occur this way. But a mutation that loses information, even if it's good, cannot be a typical member of a chain of mutations for cumulative selection. The prototype of the mutations that are supposed to make up neo-Darwinian macroevolution must be one that adds a small amount of information.[14]

Although empirical evidence shows that essentially all mutations result in a loss of information, let us consider, in accordance with the proviso for scientific integrity, the remote possibility that some mutations can result in a gain of information. If the terms "microevolution" and "macroevolution" are retained with their usual meanings, in the Cartesian Coordinate System microevolution will add cumulatively to macroevolution along either the positive y-axis (information gain) or the negative y-axis (information loss). In other words, microevolution can approach macroevolution by positive accumulation or negative accumulation. Microevolution, proceeding along the positive y-axis, will result in different levels of novelties, resulting in the formation of new organisms and increased complexity. This is the claim of the Darwinian paradigm. Microevolution, advancing along the negative y-axis, will result only in minor variants of the organism. To a limit, presumably, this is the arena for the breeding of animals and plants.

Evolution relies on the progression of time to occur; however, there is no direct correlation between level of evolution and length of time. The prefix in "microevolution," for example, could imply small-scale changes over either a short period or a long period. If the prefix "micro" is dropped, the specification of time length is avoided without affecting the level of evolution. Therefore, for more precise definitions of the two autonomous fields of evolution in terms of levels, this book replaces "microevolution" with "*intraevolution*" and "macroevolution" with "*extraevolution*." The prefix "intra" implies inside or within, and "extra" implies outside and beyond. These two terms designate the evolutionary levels without any constraints on time, and hopefully they contribute to resolving the microevolution-macroevolution controversy.

In reference to the Cartesian Coordinate System, the plane of the first quadrant above the organism's level ($y > 0$) is what I have referred to as the region of extraevolution, and the plane of the fourth quadrant at or below the organism's level ($y \leq 0$) is the region of intraevolution. While *intraevolution* and *extraevolution* are autonomous fields of evolution, *microevolution* and *macroevolution* are simply measures of evolution, just as a micrometre and metre are measures of distance. Microevolution represents small changes within or outside the organism's level *as a result of micro events* such as a minor loss/gain of information. Macroevolution, in contrast, represents comparatively large changes within or outside the organism's level *as a result of macro events* such as substantial loss/gain of information. In the field of intraevolution we can have microevolution as well as macroevolution that result in mutations involving different degrees in loss of information. In this case we will always end with variants of the same organism.

For instance, Ernst Mayr in *What Evolution Is* discusses the prolonged treatment of bacteria with penicillin, during which process an almost completely susceptible species of bacteria evolved into a totally resistant one. Since both the susceptible species and the resistant species are still bacteria, one would normally interpret this event as microevolution, but Mayr construes this event as evidence of macroevolution.[45] This is a wholly faulty conclusion. The level of evolution in his work is still micro, but because of the prolonged treatment Mayr categorizes the event as macroevolution. He therefore erroneously concludes that microevolutionary processes are simultaneously macroevolutionary processes. In order for a valid conclusion to be reached, three important scientific questions must be addressed.

First, does the change from susceptible to totally resistant bacteria represent a loss of information? If so, then instead of progressing toward novelty (upward or positive), it has regressed (downward or negative). Evolution in this range does not lead to increased complexity but simply to variants of the organism, and dependence on time is irrelevant. Second, if the change in status of the penicillin-treated bacteria involves a gain of information, there is progression toward novelty. Here reliance on time is absolute. Third, if the change involves neither loss nor gain in information, then the bacteria in question retain their status except for a reshuffling of information. Time is not of essence in this case.

Let us analyze Mayr's data in light of the newly proposed terminology. During the prolonged application of penicillin causing a loss of information, the non-resistant bacteria went from microevolution (small resistance) to macroevolution (large resistance). The final resistant species, after the application of penicillin, remain bacteria; hence, it falls under intraevolution. Macroevolution marks the limit for the penicillin treatment. At this level the species may become sterile. Thus, both microevolutionary and macroevolutionary processes can occur under intraevolution with no novelties such as new organisms and information. Thus, under the revised definition microevolution can add up to macroevolution, and hence microevolutionary processes can be simultaneously macroevolutionary processes, as Mayr asserts,[43] but as autonomous fields intraevolutionary processes (loss of information) are not simultaneously extraevolutionary processes (gain of information).

In extraevolution there would be a gain in information during the formation of transitional stages, such as part-ape/part-human, as it progresses from microevolution to macroevolution. Evolution from simple to complex, if it were possible, can only be achieved through an extraevolutionary process. Are there empirical data relating to extraevolutionary processes that support Darwinian bacteria-to-human evolution? Gary Parker has described the requirements for Darwinian evolution as follows:

> Any real evolution (macroevolution) requires an *expansion* of the gene pool, the *addition* of new genes (genons) with new information for new traits, as life is supposed to move from simple beginnings to ever more varied and complex forms ("molecules to man" or "fish to philosopher").[47]

According to this school of thought, extraevolution, which is outside the organism's level or boundary, has never been observed, and its non-existence is reflected in the abrupt gaps of the fossil record. Biological regression is what Darwinists hail as biological progression. It is not possible for blind, mindless processes such as natural selection to generate information from matter as Parker illustrates:

> The information in a book, for example, cannot be reduced to, nor derived from, the properties of the ink and paper used to write it. Similarly, the information in the genetic code cannot be reduced to, nor derived from, from the properties of matter nor the mistakes of its mutations; its message and meaning originated instead in the mind of its Maker.[48]

No real scientist doubts the role of natural selection in biological systems, but as in artificial selection its role is limited as Coffin indicates:

- Science has made an unwarranted assumption about the role of natural selection in evolution. For evolution to progress, living things must become increasingly complex and specialized. What operates through natural selection to produce such evolutionary advancement? The organism could become more fit for its environment, but does that automatically make it more complex? The dark phase of the peppered moth is truly different from the white one, and it is able to survive better in industrialized areas, but we can't claim any increased complexity for it.
- Scientists have overplayed survival of the fittest, or natural selection, as a driving force for evolution. That it does operate to some extent is clear, but its results are limited, and much change is that only—change. Neither harming nor benefiting, it does not increase complexity, nor does it lead to macroevolution (extraevolution).[49]

The mind that programmed the raw material for evolution also limits the role that natural selection can play. This is presumably the same limit as in artificial selection, which Francis Hitching mentions:

> Every series of breeding experiments that has ever taken place has established a finite limit to breeding possibilities. Genes are a strong influence for conservatism, and allow only modest change. Left to their own devices, artificially bred species usually die out (because they are sterile or less robust) or quickly revert to the norm.[50]

Natural selection can produce varieties of the same organism (bacteria-to-bacteria) but does not lead to multiple transitions from one organism to another (bacteria-to-human). Therefore, the merit of the evolutionary worldview, under the mutation-natural selection paradigm, is only in explaining the diversity of the same organism but not in transformation of one organism to another of greater complexity.

The utility of the new terminology proposed in this book is not far-fetched. With the prefixes "micro" and "macro," evolution, whose level and time dependence is complex, is presented in the form of a unit of measure; however, with the prefixes "intra" and "extra," evolution is presented or defined as a discrete process. The terms microevolution and macroevolution are vague, whereas the terms intraevolution and extraevolution are perspicuous.

Quite apart from the above, there is an inverse implication when "micro" and "macro" are used to describe evolution. Whereas in science "micro" is synonymous with small things (or small changes) that are not simple to observe, "macro" is synonymous with large things (or large changes) that are comparatively easy to observe. The reverse obtains in evolution: microevolution occurs as an observable event, but macroevolution is unobservable. It does not make sense that what is considered a small change in evolution can be recognized, while what is claimed as a large change that results in more complex structures cannot be witnessed because it requires eons in the evolutionary time frame.

In the field of intraevolution science has detected both micro and macro events that involve no new structures; thus, both microevolution and macroevolution are established scientific facts in this sense. Hence, intraevolution is a scientific fact, but in the field of extraevolution science has not observed either a micro or macro event that involves the formation of new structures or novelties, so extraevolution is not an established scientific fact. In breeding this form of evolution is forbidden, a fact that is also supported by abrupt gaps in the fossil record. Without the explicit terms intraevolution and extraevolution, all macro events, whether they lead to the formation of new structures (novelties) or not, are claimed as a scientific fact. This is scientifically incorrect! As a result, the public is confused when creationists and evolutionists engage in their usual rhetoric over whether macroevolution is a scientific fact or a myth.

Based on the evidence presented in this chapter, the following definitions are a better theoretical and experimental depiction of biological evolution.

> **Intraevolution:** Evolutionary change at or below the organism's level, either within the vicinity (microevolution) or beyond (macroevolution), that results in limited changes in appearance of populations and species over generations.
>
> **Extraevolution:** Evolutionary change above the organism's level that relates to the origins of species or the production of novelties such as new organs.

The above definitions are simply a rewording of Mayr's definitions to match the empirical evidence. The revised definitions agree with Mayr's assertions in that microevolution and macroevolution exist as scientific notions at and below the organism's level and that microevolutionary processes are simultaneously macroevolutionary processes. The revised definitions, however, rule out the view that microevolution and macroevolution are autonomous fields of biological evolution. And this is consistent with common sense, since microevolution cannot lead to macroevolution and at the same time be classified as an autonomous field. Accepting that one leads to the other justifies their classification under a common term—intraevolution.

It is doubtful that neo-Darwinians would accept the suggested changes in terminology, as doing so would preclude their claiming all aspects of evolution as a scientific fact. The current terms microevolution and macroevolution were chosen solely on grounds of convenience and were applied to biological evolution in order to insinuate equality between the mechanisms of macro—and microevolution. This is no longer tenable since the empirical evidence consistently suggests otherwise.

It is worthwhile to compare evolution that is scientific fact (intraevolution) and evolution that is simply a philosophical opinion (extraevolution) and, thus, a belief that parallels belief in creation (see comparative details in Table 3.1).

Table 3:1 Comparison between the Scientific Aspect of Evolution (Intraevolution) and the Dogmatic Aspect (Extraevolution).

Intraevolution [mutation—selection]	Extraevolution [pseudoscience]
Changes in appearance of species. Evolution below and at the species level. The truth is in the evidence of different kinds of cats, dogs, etc.	Origins of species. Evolution above the species level. The fallacy is the absence of the various stages of transformations from molecules to human.
No bizarre anti-scientific assumptions, such as that life emanates from non-life, are required. Creationists and evolutionists are in agreement.	Assumes that life originated from non-life, scientifically proven false and contrary to our experience. Creationists and evolutionists split on philosophical grounds.

Can be falsified, hence constitutes regular science.	Cannot be falsified. Proponents contend it requires eons to happen; hence, it falls short of being classified as regular science.
Scientists are unanimous about the mechanisms involved, which are a combination of environmental changes or adaptation, mutation, and selection.	Mechanisms unknown because they are rooted in a false assumption. Constant battle rages among leading evolutionists.
Has medical, agricultural, and industrial applications. Indispensable to the study of biology.	Has no technological applications since it is a belief. Has religious and social implications. Dispensable for the study of biology.

Therefore, any extrapolation from intraevolution to extraevolution is an extrapolation from science to pseudoscience, from fact to myth. The presupposition that anything about creation is religion and everything about evolution is science is, therefore, incorrect. The discord within the scientific community is only over extraevolution.

In sum, there is ample evidence that the evolutionary paradigm does not comply with the ground rules for scientific integrity. At present microevolution and macroevolution are treated as autonomous fields as well as measures of the level or magnitude of evolution. This is a recipe for misunderstanding and has been a constant source of friction between creationists and evolutionists. I addressed this matter in my previous book, *The Death of Evolution*.[51] I now am suggesting that, in order to resolve a common misunderstanding, the prefix "intra" should replace "micro" in describing evolution within the species boundary and "extra" should replace "macro" for evolution outside the species boundary. The rationale is that "micro" and "macro" refer to both levels of evolution as well as the length of time involved. No one knows how these parameters are related. Therefore, a micro-level event over a long period may be misconstrued as a macro-level event. In addition, to complicate things, the evolutionary process is alleged to be gradual and sometimes in punctuated equilibrium. Dropping the prefixes from microevolution and macroevolution allows the level of evolution to be interpreted without the time ambiguity.

Intraevolution is a scientific fact, and all productive applications of evolutionary concepts are confined to this autonomous field of evolution. Extraevolution is a belief in nonexistent events and thus cargo-cult science (pseudoscience).

NOTES

1. "The Second Evolution: The Unified Teem Theory of Evolution, Perception, Emotions, Behaviour and Inheritance," http://thesecondevolution.com/Introduction.pdf. Retrieved on 3 April 2011.

2. Glossary, http://www.pbs.org/wgbh/evolution/library/glossary/glossary.html. Retrieved on 3 April 2011.

3. *The Complete Idiot's Guide to Evolution* (Indianapolis: Alpha Books, 2002), p. 181.

4. Understanding the Creation-Evolution Controversy (Ozark, AL: ACW Press, 2005), p. 25.

5. "Microevolution vs. Macroevolution: Is There a Difference between Microevolution and Macroevolution?" http://atheism.about.com/od/evolutionexplained/a/micro_macro.htm. Retrieved on 16 December 2008.

6. *Dawkins Vs. Gould: Survival of the Fittest* (Thriplow, Cambridge: Icon Books, 2007), p. 178.

7. *The Natural Sciences Know Nothing of Evolution* (Costa Mesa, CO: T. W. F. T. Publishers, 198), pp. 135-36.

8. "Cargo[-]Cult Science," Surely You're Joking, Mr.Feynman!: Adventures of a Curious Character (New York: W. W. Norton, 1985), p.343.

9. The Evolution of Living Organisms: Evidence for a New Theory of Transformation (New York: Academic Press, 1977), p. 7.

10. Union of Concerned Scientists, "Section 4: Why Intelligent Design Is Not Science." http://www.ucsusa.org/scientific_integrity/what_you_can_do/why-intelligent-design-is-not.html. Retrieved on 22 February 2009.

11. *Evolution: A Theory in Crisis* (Bethesda, MD: Adler and Adler, 1986), p. 65.

12. *What Evolution Is* (New York: Basic Books, 2001), p. 188.

13. "Is Evolution a Theory, a Fact, or a Law?" *St. Louis Metro Voice,* October 1993. http://emporium.turnpike.net/C/cs/theory.htm. Retrieved on 3 April 2011.

14. Of Pandas and People: The Central Question of Biological Origins (Dallas: Haughton Publishing, 1989, pp. 10-11.

15. Evolution: A Theory in Crisis, pp. 64-65.
16. Gary Parker, *Creation: Facts of Life* (Green Forest, AR: Master Books, 2006), pp. 145, 151.
17. Theodosius Dobzhansky, "On Methods of Evolutionary Biology and Anthropology," *American Scientist* 45 (1957), p. 385.
18. Of Pandas and People, p.11.
19. Ibid.
20. R. B. Goldschmidt, "Evolution, as Viewed by One Geneticist," *American Scientist* 40 (1952), p. 86.
21. Henry M. Morris, *Scientific Creationism* (Green Forest, AR: Master Books, 1998), p. 55.
22. H. J. Muller, "Radiation Damage to the Genetic Material," *American Scientist* 38 (1950), p. 35; Morris, *Scientific Creationism*, p. 55.
23. "On Methods of Evolutionary Biology and Anthropology," p. 385.
24. *The Problems of Evolution* (New York: Oxford University Press, 1985), p. 121.
25. Scientific Creationism, p. 57.
26. Of Pandas and People, pp. 11-12.
27. Ibid, p. 12.
28. *Creation and Change* (Fearn, Ross-Shire: Mentor, 1999), p. 198.
29. Ibid.
30. Of Pandas and People, p.12.
31. *Darwin and Intelligent Design* (Minneapolis: Fortress Press, 2006), p. 64.
32. Creation: Facts of Life, p. 120.
33. "Adaptation," *Scientific American*, September 1978, p. 215.
34. Creation: Facts of Life, p. 96.
35. *Genetics and the Origin of Species* (New York: Columbia University Press, 1951), p. 73.
36. Creation: Facts of Life, pp. 111-12.
37. *Origin by Design* (Hagerstown, MD: Review and Herald Publishing, 1983), p. 407.
38. *The Neck of the Giraffe* (London: Pan Books, 1982), p. 57.

39. *Tornado in a Junkyard* (Arlington, MA: Refuge Books, 1999), p. 50.
40. Not by Chance!: Shattering the Modern Theory of Evolution (New York: Judaica Press, 1998), p. 138.
41. Ibid, p.141.
42. Ibid, pp. 141, 143.
43. Ibid, p. 143.
44. Ibid, p. 148.
45. What Evolution Is, p. 190.
46. Ibid.
47. Creation: Facts of Life, p. 133.
48. Ibid, p. 122.
49. Origin by Design, p. 408.
50. The Neck of the Giraffe, p. 55.
51. The Death of Evolution: God's Creation Patent and Seal (Longwood, FL: Xulon Press, 2007), p. 227.

CHAPTER 4

CREATION VS. EVOLUTION COSMOLOGY

Belief in the theory of evolution is exactly parallel to belief in special creation—both are concepts which believers know to be true but neither, up to the present, has been capable of proof.[1]

—L. Harrison Matthews

I am an ardent evolutionist and an ex-Christian, but I must admit that . . . evolution is a religion. This was true of evolution in the beginning, and it is true of evolution still today . . . . Evolution therefore came into being as a kind of secular ideology, an explicit substitute for Christianity.[2]

—Michael Ruse

Both the evolutionary materialist and the fundamentalist "creation scientist" are quite alike in their contaminating aspects of pure science with large doses of doctrine, though the beliefs are different in each case.[3]

—John F. Haught

"Scientific materialism resembles religion," contends John F. Haught, "and can be called a belief system, because it systematically answers many of the same ultimate questions that religion responds to: Where do we come from? Where are we going? What is the deepest nature of reality? What is our true identity? Is there anything permanent and imperishable? etc. The answer to all of these questions, according to scientific materialism, centres on the concept of 'matter.' The clarity and simplicity—the hard-rock 'realism'—of the idea of 'matter' has enormous appeal to many scientists and philosophers. It satisfies a deeply religious longing for a solid and comprehensive ground upon which to base their knowing and being."[4]

Evolutionism claims to reveal where we came from, who we are, and what our destiny is. These are values that shape our perception of the world. A student, therefore, has the right to ask any question relevant to human origin, whether in a scientific or religious forum. A teacher has the moral obligation to respond, irrespective of the forum, in a manner that involves inclusion and not exclusion, in an atmosphere that nurtures fairness and not intimidation, and in a way that welcomes reason and not its denial.

We must, as Haught points out, "consistently and rigorously distinguish science from all belief systems, whether religious or materialist."[5] Evolution has a scientific component (intraevolution) that is distinguishable from a materialist or religious component (extraevolution). In this chapter we will examine both the creationist and evolutionist worldviews in light of these terms.

CREATIONIST VS. EVOLUTIONIST WORLDVIEWS

Creation and evolution are facts of life. However, in terms of worldviews concerning life's origin, the topic of creation is almost synonymous with religion and hence myth. Evolution, on the other hand, is synonymous with science and thus fact. In reality, though, creation and evolution are both generic terms, and each encompasses two fields—one scientific and the other dogmatic. In science, for instance, we can talk about the creation of a drug or the atomic bomb. These are processes that, undertaken by scientists in a laboratory, can be repeated at any time and whose results are reproducible in other laboratories. No supernatural revelation is required. There is also the dogma of creation or special creation that concerns the origins of the universe and species, which were not undertaken by scientists and, consequently, cannot be repeated. These events did not depend on chance and were not witnessed by any human being; instead, they were revealed or reported by the inventor. The validity of these revelations can be verified by the use of scientific models. This is what constitutes creationism, which is not a branch of regular science.

William Stoeger has addressed use of the word "creation" from both a scientific and philosophical point of view:

> In science the word "creation" is used rather loosely. Whenever a new particle appears as the result of an interaction of other particles or by polarization of the vacuum (vacuum fluctuations), we speak of its creation. In other contexts we speak of stars being created out of collapsing and fragmenting clouds of hot nebular gas, of proteins being created out of amino acids according to the genetic blueprints carried by messenger RNA, of a zygote being created from the fusion of ovum and sperm. But, from a philosophical point of view . . . we mean something much narrower and more specific by "creation." We mean creation *ex nihilo*, both the *ultimate* bringing into, and maintaining in, existence—creation out of absolutely nothing. The creation with which we are familiar in the sciences . . . is not creation *ex nihilo*, but rather the transformation of previously existing material or physically accessible entities into qualitatively new ones.[6]

In this book "dogmatic creation" is what Stoeger refers to as creation *ex nihilo*. Gary Parker uses the DNA molecule to provide a striking analogy between creation by human hand and that found in living systems:

All of us can recognize objects that man has created, whether paintings, sculptures, or just a Coke bottle. Because the pattern of relationships in those objects is contrary to relationships that time, chance, and natural physical processes would produce, we know an outside creative agent was involved. I began to see the same thing in a study of living things, especially in the area of my major interest, molecular biology. All living things depend upon a working relationship between inheritable nuclei acid molecules, like DNA, and proteins, the chief structural and functional molecules. To make proteins, living creatures use a sequence of DNA bases to line up a sequence of amino acid R-groups. But the normal reactions between DNA and proteins are the "wrong" ones, and act with time and chance to disrupt living systems. Just as phosphorus, glass, and copper will work together in a television set only if properly arranged by human engineers (as outside creative agent), so DNA and protein will work in productive harmony only if properly ordered by an outside creative agent. I presented the biochemical details of this DNA-protein argument to a group of graduate students and professors, including my professor of molecular biology. At the end of the talk, my professor offered no criticism of the biology or biochemistry I had presented. She just said that she didn't believe it because she didn't believe there was anything out there to create life.[7]

Notice in this anecdote that, because evolutionism is pseudoscience, a brilliant scientific question from a student gets a shallow philosophical answer from his professor. Parker provides an excellent illustration of how DNA is the evidence of intelligence as opposed to non-intelligence. The professor's response to Parker confirms that evolutionism is driven by atheistic beliefs rather than objective science. It affirms atheist Richard Lewontin's vow that modern evolutionists cannot allow a "Divine Foot in the door."[8] Parker's argument is one of the reasons why evolutionists fight to exclude creationist's ideas in science classrooms. Ironically, some advocates of dogmatic evolution, such as Richard Dawkins, believe in aliens or god-like extraterrestrials.[9]

The field of evolution that deals with the origins of anatomically and genetically different organisms is appropriately referred to as extraevolution. The validity of extraevolutionary hypotheses can be examined in light of scientific models. This is what constitutes evolutionism, which is not a branch of regular science. L. C. Birch and P. R. Ehrlick describe the attributes of dogmatic evolutionism as follows:

> Our theory of evolution has become . . . one which cannot be refuted by any possible observations. Every conceivable observation can be fitted into it. It is thus "outside empirical science" but not necessarily false. No one can think of ways in which to test it. Ideas, either without basis or based on a few laboratory experiments carried out in extremely simplified systems, have attained currency far beyond their validity. They have become part of an evolutionary dogma accepted by most of us as part of our training. The cure seems to us not to be a discarding of the modern synthesis of evolutionary theory, but more skepticism about many of its tenets.[10]

This confirms that extraevolution lies outside the parameters of empirical science and thus cannot be tested or falsified. Dogmatic creationism (creation *ex nihilo*) parallels dogmatic evolutionism (evolution *ex nihilo*) in this regard.

Some assume that biological evolution is simply a theory and not a fact, but some think otherwise. In a 1999 booklet NASIM defines scientific "fact" and "theory" as follows:

Fact: In science, an observation that has been repeatedly confirmed and for all practical purposes is accepted as "true." Truth in science, however, is never final, and what is accepted as

a fact today may be modified or even discarded
tomorrow.

Theory: In science, a well-substantiated explanation
of some aspect of the natural world that
can incorporate facts, laws, inferences, and
tested hypotheses.[11]

Here some further explanation is necessary. A scientific theory
can either be correct (scientific knowledge) or wrong (scientific
myth). A scientific theory may incorporate facts and laws of
nature, but it is not itself necessarily a scientific fact/knowledge.
For instance, the neo Darwinian theory of evolution incorporates
facts about species from fossils, DNA, mutation, etc. This does not
make the Darwinian explanation of evolution a comprehensive
scientific fact. To qualify as a scientific fact, a prediction or
observation must be tested empirically and retested to confirm
it. In this sense bacteria-to-bacteria evolution is a scientific fact,
but bacteria-to-human evolution is not.

Ardent evolutionists regard evolution as much more than a
theory. Pierre Teilhard de Chardin, a Jesuit priest and paleontologist,
describes dogmatic evolutionism (extraevolution) in terms of
deific qualities:

> For many, evolution is still only transformism, and
> transformism is only an old Darwinian hypothesis
> as local and as dated as Laplace's conception of the
> solar system or Wegener's Theory of Continental Drift.
> Blind indeed are those who do not see the sweep of a
> movement whose orbit infinitely transcends the natural
> sciences and has successively invaded and conquered
> the surrounding territory—chemistry, physics, sociology
> and even mathematics and the history of religions. One
> after the other all the fields of human knowledge have
> been shaken and carried away by the same under-water
> current in the direction of the study of some *development*.
> Is evolution a theory, a system, or a hypothesis? It is much
> more: It is a general condition to which all theories, all

systems, all hypotheses must bow and which they must satisfy henceforward if they are to be thinkable and true. Evolution is a light illuminating all facts, a curve that all lines must follow.[12]

According to this statement, evolution is an almighty "condition" to which all knowledge must bow. It transcends both religion *and* science. The virtues of science are relinquished as evolutionism overrules some of its laws. For instance, it is a scientific law that life cannot be generated spontaneously from dead matter. In order to propagate their views, however, evolutionists argue without proof that under the right conditions life can be generated from non-life. Thus, from a scientific standpoint extraevolution begins with a false or delusory premise. By violating established empirical knowledge, *extraevolution cannot be deemed a scientific fact*. This is strong evidence that evolutionists force scientific data to fit their theory. While true science prides itself on openness to conceptual change, evolutionism as a religion is a fixed dogma.

It is evident from Table 4.1 below that the modern world is faced with two choices: a creationist worldview that acknowledges a supernatural designer, or an evolutionist worldview that privileges the mindless process of natural selection. It comes down to a choice between a conscious designer and an unconscious instrumentality. If natural selection is the correct choice, it has two weaknesses. First, it depends upon a superior and independent agency for the raw materials it tinkers with, since it does not create them. Second, as a subordinate mechanism, natural selection, like the origin of life, needs an explanation. In essence, natural selection owes its existence to a supernatural agency. The ultimate question is whether our choice should be based on philosophical preference or empirical reality. In opting for the former over the latter, ironically, many leading scientists today, unlike their predecessors Sir Isaac Newton and Albert Einstein, have chosen natural selection over supernatural intelligence. It is a philosophical issue, and scientists are divided.

Table 4:1 A Comparison of Special Creation (Creationism) and Special Evolution (Evolutionism)

Creationism	Evolutionism
God is all-powerful. BIG WORDS The omniscient, omnipresent, and omnipotent God framed, ordered, and made everything from nothing in this world. (Evidence of supernatural power, understanding, wisdom, and plan.)	Time and chance are all-powerful. BIG BANG Concentrated matter and energy accidentally exploded into everything in the universe at some point in the cosmological past. (Natural miracle happening by accident and chance.)
ORGANISMS CREATED BY UNDERSTANDING Fully formed organisms were made as unique species and biologically programmed to reproduce after their kind, subject to mutational degradation and natural selection. (Life originated from God. This is consistent with the scientific law of biogenesis. The fossil record confirms what exists in the living world. Organisms appear fully formed with abrupt gaps that affirm no transitions between organisms.)	ORGANISMS EVOLVED BY CHANCE Organisms evolved spontaneously from dead matter and progressed from simple to complex forms by random genetic mutations and natural selection. The source of genetic material and the origin of genetic information are unknown. (The hypothesis of spontaneous generation violates the scientific law of biogenesis and common sense. The fossil record does not confirm the many transitional links envisaged. Every claim for the discovery of missing links is disputable.)

HUMAN BEINGS MADE IN THE IMAGE OF GOD	HUMAN BEINGS MADE IN THE IMAGE OF NOTHING
God is the source of consciousness, moral values, spirituality, and abstract or analytical intelligence.	The origin of consciousness, moral virtues, spirituality, and abstract reasoning is unknown.
CREATION WITHIN SIX DAYS	EVOLUTION OVER MILLIONS OF YEARS
The task of designing the earth, its environment, and its occupants lasted six days. (Creation in six days is evidence of power, wisdom, and understanding. It is not a measure of the earth's age, just as the structural age of a building is not a measure of the age of its components.)	Organisms evolved from bacteria by random mutation and natural selection over millions of years. (This theory is based on a speculative projection about the unknown past and limited scientific knowledge. In short, it is pure conjecture.)
SIMILARITIES AMONG ORGANISMS	SIMILARITIES AMONG ORGANISMS
God used similar materials and style in designing both the DNA and the morphologies of various organisms.	Organisms descended mutationally from a common primordial ancestor.
DISSIMILARITIES AMONG ORGANISMS	DISSIMILARITIES AMONG ORGANISMS
God endowed the various organisms with different non-material virtues (different minds, expression of love, moral values, knowledge, soul, spirit, etc.).	A mindless mechanism cannot generate different minds or intelligent beings with spiritual values. This is evidence that various organisms did not derive from a common primordial

	ancestor but from radically different primordial ancestors.
(This explains why organisms are both similar and dissimilar in many respects.)	(This explains why organisms are similar but not why they are dissimilar. Materialism cannot account for non-material virtues.)

The conflict within the scientific community is not over science but over the philosophical interpretation of the same scientific evidence. Ariel A. Roth writes:

> Both evolutionists and creationists accept the data of science but place different understandings on them. For instance, evolutionists teach that the similarities in cell structure, biochemistry, and anatomy found among different kinds of animals and plants are the result of a common evolutionary origin, whereas creationists look at the same data and interpret them as representing the imprint of a single designer, God . . . . Open conflict between scientific interpretations and the Bible has raged for two centuries. It is one of the greatest intellectual battles of all time. The instruments of battle are the pen and tongue, and the battlefield is the mind. This question affects our basic worldview, our reason for existence, and our hope for the future. It is an issue that we cannot easily lay aside.[13]

Leading modern scientists have chosen natural selection over a supernatural Creator for two reasons: materialism and the principle of Occam's razor.

The latter is used as a philosophical guide to choosing between two plausible theories. In science the principle suggests simply that "when two competing theories make equally valid predictions or explanations, choose the simpler."[14] So, in addressing the creationism-evolutionism controversy, Dylan Evans and Howard Selina tender the following argument:

> [T]here are two possible explanations for complex designs. Either they were designed by God, or by cumulative natural selection. Which explanation should we prefer? Whenever there is more than one possible explanation for something, there is a simple rule of thumb that we should use to decide which is the right explanation . . . . ALWAYS CHOOSE THE SIMPLEST EXPLANATION. This rule is called "Occam's razor, after William of Occam (c. 1285-1347). Natural selection is a much simpler explanation than divine creation because it only requires us to believe in things that we already know about . . . . Thus we should always prefer to explain biological design by natural selection rather than by divine creation.[15]

Is this a valid argument? Natural selection works only on raw materials that are already complex. Since complexity precedes natural selection, how can natural selection be a possible explanation of complex designs? As such, natural selection offers no simpler explanation but is just a preferred choice to justify the preconceived materialistic explanation of evolutionists. We know natural selection primarily in its capacity to sustain adaptation. And we do not know natural selection any better than we know artificial selection. Despite all its elegance, artificial selection has not produced any transitional organism.

The claim that natural selection creates novelty is pure illusion. Nobel laureate Ernst Boris Chain reminds the public: "Actually, scientists are often just as prejudiced in their theories and emotionally involved in the implications of their work as are other nonscientific members of society, and are unreliable in their predictions and interpretations."[16] Concerning the principle of Occam's razor, another Nobel laureate, Francis Crick, cautions: "While Occam's razor is a useful tool in the physical sciences, it can be a very dangerous implement in biology. It is thus very rash to use simplicity and elegance as a guide in biological sciences."[17]

Biologists really do not know the mechanism of evolution by natural selection. Philip E. Johnson affirms Crick's statement:

> The biologists are at each other's throats in private, fighting over every detail in the Darwinist scientific program. The versions of "evolution" promulgated by Richard Dawkins and Stephen Jay Gould, for example, have hardly anything in common except their common adherence to philosophical materialism and their mutual dislike for supernatural creation.[18]

Although biologists may be fighting among themselves behind closed doors, their "mutual dislike for supernatural creation" nonetheless prompts them to put up a publicly united front in advocating evolution as a scientific fact.

Leading evolutionist Francisco J. Ayala presumes to speak on behalf of the entire scientific community when he asserts:

> Scientists agree that the evolutionary origin of animals and plants is a scientific conclusion beyond reasonable doubt. They place it beside such established concepts as the roundness of the earth, its rotation around the sun, and the molecular composition of matter. That evolution has occurred, in other words, is a fact.[19]

There are five reasons, however, why Ayala's claims cannot be justified. I will discuss the first three together because of their affinity.

First, the phrase "evolutionary origin" is ambiguous and misleading. Evolutionists must prove this assertion. So far there is no hard scientific evidence to support it. Second, there are almost as many evolution fundamentalists as there are creation fundamentalists. All scientists accept evolution as a process of change in microevolution (intraevolution) but not as the origin of species claimed in macroevolution (extraevolution). Since scientists disagree only on philosophical grounds, the number of advocates on either side of the controversy is not crucial from a scientific point of view. Third, the story of origins, like the mystery of love, is one of the gaps in our phenomenal world that science cannot explain. Science is about repeatability and consistency. The bacteria-to-human model of Darwinian evolution cannot be demonstrated because it is not repeatable.

The circumstantial evidence that evolutionists cite to justify evolution as a fact is the same that creationists use more convincingly to explain creation as a fact. For example, evolutionists claim that similarity in morphology and DNA is strong evidence for a "common ancestry" of all life forms. They provide a complex evolutionary tree to support their choice. Since God used clay to create all life forms, creationists employ the "common plan and design" of a creator to explain the same evidence. As ex-evolutionist Gary Parker points out, "At the atomic level ('dust of the ground'), all organisms are essentially 100 percent identical."[20] Therefore, both arguments are valid. However, when we incorporate other pieces of evidence, the argument for the "common designer" is simpler and more convincing than the evidence of common ancestry. Ayala would be more convincing if he proclaimed creation rather than evolution as a fact.

As far as the origin of organisms is concerned, biologists are one-sided in their studies. How can they faithfully follow the scientific evidence? Little wonder that evolutionists, in order to promote evolutionism, sometimes mistakenly explain common design as common ancestry.

Tim Berra explains evolutionary diversity and complexity by means of the following illustration:

> Everything evolves, in the sense of "descent with modification," whether it be government policy, religion, sports cars, or organisms. The revolutionary fiberglass Corvette evolved from more mundane automotive ancestors in 1953. Other high points in the Corvette's evolutionary refinement included the 1962 model, in which the original 102-inch was shortened to 98 inches and the new closed-coupe Stingray model was introduced; the 1968 model, the forerunner of today's

Corvette morphology, which emerged with removable
roof panels; and the 1978 silver anniversary model, with
fastback styling. Today's version continues the stepwise
refinements that have been accumulating since 1953. The
point is that the Corvette evolved through a selection
process acting on variations that resulted in a series of
translational forms and an endpoint rather distinct from
the starting point. A similar process shapes the evolution
of organisms.[21]

In this comparison cars, the equivalent of organisms, have similar
body parts, composition, and engines. The different models of cars
do not evolve by random chance or mindless processes. Instead,
they are purposely designed. Phillip E. Johnson, commenting
on Berra's explanation of evolution by analogy to the Corvette's
production, writes:

It illustrates how intelligent designers will typically
achieve their purposes by adding variations to a basic
design plan. Above all, such sequences have no tendency
whatever to support the claim that there is no need for
a Creator, since blind natural forces can do the creating.
On the contrary, they show that what biologists present
as proof of "evolution" or "common ancestry" is just as
likely to be evidence of common design.[22]

Arguments based solely on similarity are inadequate to the
important challenge of reaching valid conclusions about Darwinian
evolution.

How, then, do we account for the variety of life forms? At
creation God mandated that species should reproduce after their
kind, multiply, and fill the earth (Gen. 1:11-22, 28). For most
organisms this process begins when sperm fertilized an ovum and
its DNA combined with that of the ovum. By way of comparison,
evolutionists claim that it took millions of years to develop from
molecular to human complexity. Under the control of natural
processes that God has set in motion, however, this transformation

from cell (zygote) to baby is accomplished within the womb in nine months. We therefore must ask ourselves, "Is the process of reproduction/embryonic development, which God instituted at creation, an evolutionary or creative process?" There are two schools of thought on this issue.

Representing the creationist camp, Phillip E. Johnson asserts in *Defeating Darwinism*: "Embryonic development is a programmed process that proceeds directly to a preordained end point. The apparent impossibility of using chance mutations to alter embryonic development so as to produce a different kind of animal agues strongly against the claims to Darwinian macroevolution."[23] In contrast, evolutionist Leslie Alan Horvitz writes: "According to Ernst Haeckel, a nineteenth-century German biologist and philosopher, 'Ontogeny recapitulates phylogeny'—fancy words meaning that the development of the human embryo retraces the evolutionary development of the human race, each beginning with a simple organism (a cell) and evolving into complex forms."[24] Evolutionists cannot have it both ways. On the one hand, they proclaim evolution as an aimless, random process involving millions of years; on the other hand, they declare embryonic development, which is a planned, purposeful process accomplished in less than a year.

The fourth point concerns Ayala's comparison of the facts of evolution with matter's molecular composition. Because the latter does not tell us about the origin of atoms, molecules, or matter, this is not a valid comparison. Just as scientists cannot explain the origin of life, so they are not able to demonstrate the origin of species

Fifth, evolution is a fact as a process of change but not of origins. With regard to the origins of living forms, evolution is at best a hypothesis rather than a theory. Chain thus writes:

> It is, of course, nothing but a truism, and not a scientific theory, to say that living systems do not survive if they are not fit to survive. It is equally obvious that if in a given species chance mutations give rise to genetic variants which in a particular ecological surrounding have a better chance of survival than other members of the species,

this will give them an advantage in survival value over their less fit competitors. To postulate, as the positivists of the end of the last century and their followers here have done, that the development and survival of the fittest is *entirely* a consequence of chance mutations, or even that nature carries out experiments by trial and error through mutations in order to create living systems better fitted to survive, seems to me a hypothesis based on no evidence and irreconcilable with the facts. This hypothesis willfully neglects the principle of teleological purpose which stares the biologist in the face wherever he looks, whether he be engaged in the study of different organs in one organism, or even of different subcellular compartments in relation to each other in a single cell, or whether he studies the interrelation and interactions of various species.[25]

Clearly biologists have ignored, or used hand-waving arguments to explain away, the inherent complexities involved in promoting the concept of natural selection as the origin of species. Former evolutionist Gary Parker points out this problem:

Evolution demands an increase in the quantity and quality of genetic information, and mutation-selection, *no matter how long you wait*, cannot provide it. *But both mutation and selection are* very real, observable processes going on around us every day. *Evolution,* **no,** *but mutation-selection,* **yes**! They don't produce evolutionary changes, but mutation and selection do indeed produce changes. Mutations are no real help in explaining the origin of *species*, but they are great for explaining the origin of disease, disease organisms, and birth defects. Natural selection is no real help in explaining the *origin* of really new species, but it's great for explaining *how* and *where* different specialized sub-types of the various created kinds "multiplied and filled the earth" after death corrupted the creation and, again, after the Flood.[26]

It makes more sense to relate mutations, which are copying mistakes, to the origin of human diseases as opposed to the origin of human beings.

In claiming that evolutionism is a fact, scientists have based their arguments on the similarities among organisms, but organisms are characterized by both similarities and dissimilarities. With creationism the similarities and dissimilarities can be explained fully by the postulate of a designer. With evolutionism, on the other hand, the premise of a common ancestor means that the marked dissimilarities cannot be explained satisfactorily. For instance, humans and chimps may have 98% similarity in their DNA, but the marked dissimilarities outweigh their similarities. Accordingly, Chain offers these two remarks:

- We do not need to be expert zoologists, anatomists or physiologists to recognize that there exist some similarities between apes and man, but surely we are much more interested in the differences than the similarities. Apes, after all, unlike man, have not produced great prophets, philosophers, mathematicians, writers, poets, composers, painters and scientists. They are not inspired by the divine spark which manifests itself so evidently in the spiritual creation of man and which differentiates man from animals.

- Why this should be so is one of the many great and insoluble mysteries of nature, and it is idle to search for explanations of the unexplainable. To say that man has left the apes behind in the evolutionary scale because he managed, for various reasons, to develop a bigger brain is really no explanation at all; it is only a statement covering up ignorance by an ill-defined term.[27]

These dissimilarities are not programmed into our DNA, so scientists must be concerned with attributes that markedly separate apes from human beings. Tim Berra's illustration of descent with modification also shows that it is not the similarity that counts but its organization. The Genesis account of primal creation contends that human beings are made in the image of God, a claim that explains our spirituality, abstract intelligence, and moral values. Can natural selection account for all of these attributes?

Atheist Danny Vendramini argues that the only way to inherit any traits is through protein-coding genes. He also questions how our interest in mythic monsters might be related to our chromosomal inheritance, reaching the conclusion that "our predisposition to myths was not created by natural selection as there is a limit to what the random toss of genetic mutations could achieve."[28] Vendramini proposes a second evolutionary process that addresses the inheritance of innate behavior, instincts, and emotions in multicellular animals. How his theory explains the great dissimilarities among different organisms remains to be seen.

Evolutionism is therefore more a myth than a scientific fact. Michael Denton reaches a similar conclusion about Darwinian evolution:

> Considering its historic significance and the social and moral transformation it caused in western thought, one might have hoped that Darwinian theory was capable of a complete, comprehensive and entirely plausible explanation for all biological phenomena from the origin of life on through all its diverse manifestations up to, and including, the intellect of man. That it is neither fully plausible, nor comprehensive, is deeply troubling. One might have expected that a theory of such cardinal importance, a theory that literally changed the world, would have been something more than metaphysics, something more than a myth.[29]

Why, then, does the scientific community continue to promote the theory of evolution as a fact? There are three categories of scientists: first, the evolutionists, many of whom are atheists or agnostics; second, the creationists; and, third, the conformists or secular scientists. Materialism is the god of most leading scientists who speak out on behalf of the scientific community. According to Tom Bethell, the first group of apologists is prepared to defend evolutionism to the bitter end.[30]

Evolutionists have compromised scientific integrity in their desire to explain our complex and intelligently designed bodies as the product of eons-long and mindless natural selection. Neo-Darwinian advocates suggest that we have to wait millions of years for drastic transitional changes, but true science is built on empirical evidence, not delayed proof or unjustified extrapolations. These limitations disqualify evolutionism as a scientific theory. Quite apart from that, the neo-Darwinian theory of evolution presupposes the spontaneous generation of life from inorganic (dead) matter and, hence, contradicts the fundamental law of biogenesis. It also assumes the spontaneous emergence of information from matter and negates the basic premise that "information cannot come from random variations, only from pre-existing information."[31]

WHY CREATIONISM MATTERS IN A SCIENCE CLASS THAT TEACHES EVOLUTIONISM

Michael Shermer, in the appendix of his book *Why Darwin Matters: The Case Against Intelligent Design*, comments:

> Over the past century, nearly every court case and curriculum dispute in the evolution-creation debate has included some form of the "equal time" argument. Well, even if we all agreed public school science classes should spend equal time on each perspective, we must ask, Equal time for whom? My friend and colleague Eugenie Scott, Executive Director of the National Center for Science Education, outlines at least eight different positions one might take on the creation-evolution continuum.[32]

The different positions that Shermer lists include Young-Earth Creationists, Old-Earth Creationists, Gap Creationists, Day-Age Creationists, Progressive Creationists, Intelligent-Design Creationists, Evolutionary Creationists, and Theistic Evolutionists. The various categories reflect different opinions about the mechanics and time involved in Creation narratives. Shermer capitalizes on these differences to disqualify the inclusion of intelligent design in science curricula. "Note," he writes, "that the Intelligent Design creationists are but one of many competing for space in the curriculum . . . . If the government were to force teachers to grant equal time for them, then why not these others?"[33] Shermer is right if we address the creationism-evolutionism controversy based solely on the Creation stories in Genesis 1 and 2.

This book, however, addresses the debate from the standpoint of God's claim before an audience to have created the universe as documented in Exodus 19, 20, and 31. There is a remarkable difference between a claim for something and a story about something. A credible claim is subject to only one interpretation, whereas a story may have multiple interpretations. While the Genesis 1 and 2 stories about Creation are subject to different interpretations by scholars, Exodus 19, 20, and 31 recounting God's claim of ownership of the universe has just one interpretation. It is either true or false. With God's personal claim to have created the universe as the focal point, the debate about mechanics and the earth's age are no longer relevant; consequently, the divergent views of various creationist camps are rendered unimportant.

Evolutionism logically should be compared to creationism as an earlier worldview. Evolutionist Richard D. Alexander recognizes the need to present both worldviews:

> No teacher should be dismayed at efforts to present creation as an alternative to evolution in biology courses; indeed, at this moment creation is the only alternative to evolution. Not only is this worth mentioning, but also a comparison of the alternatives can be an excellent exercise in logic and reason. Our primary goal as educators should be to teach students to think[,] and such a comparison, particularly because it concerns an issue in which many have special interests or are even emotionally involved, may accomplish that purpose better than most others . . . .
> In the sense that creation is an alternative to evolution

for any specific question, a case against creation is a
case for evolution and vice versa.[34]

Any insistence that science lessons must be confined to
evolutionism exclusively is an admission of insecurity. A scientific
fact must be compared with its rational alternative. The facts/
tenets of evolutionism should be compared against the facts/
tenets of creationism, whether in a religion class or a science
class.

Admittedly, there are many creation stories in world history, but
only one Western worldview parallels the dominant evolutionist
worldview. The Judeo-Christian God associated with the "Spoken
Edict Creation Stories" fulfills this requirement as documented in
Exodus 19 and 20. Other creation stories are therefore unimportant
in the creationism-evolutionism controversy. Evolutionists craftily
bring them up as an argument to support why the creationist
worldview should not be presented side by side with the evolutionist
worldview. Presumably they fear that evolutionism might crumble
when juxtaposed to biblical creationism.

It is sheer delusion to contend that evolution "transcends the
natural sciences and has successively invaded and conquered the
surrounding territory—chemistry, physics, sociology, and even
mathematics and the history of religions."[35] Since the spontaneous
generation of living organisms from non-living matter (abiogenesis)
defies the basic laws of science, evolution by natural selection,
as the sole reason for our existence and diversity of the living
world, is simply a form of pseudoscience. It thus is preposterous
to describe evolution as "a general condition to which all theories,
all systems, all hypotheses must bow and which they must satisfy
henceforward if they are to be thinkable and true."[36]

Darwinian evolution by natural selection springs from prior
knowledge of artificial selection in plant—and animal-breeding
experiments. Therefore, there is nothing "original" or "outstanding"
about either the Darwinian or neo-Darwinian theory of evolution.
In artificial selection, in which, people rather than the environment
select for desirable traits, the different breeds do not develop
complex features. Therefore, the arguments that cumulative natural

selection is the only force that can produce complex designs like eyes or hearts is a farce.

I conclude with the following comments by H. S. Lipson:

- I have always been slightly suspicious of the theory of evolution because of its ability to account for *any* property of living beings (the long neck of the giraffe, for example). I have therefore tried to see whether biological discoveries over the last thirty years or so fit in with Darwin's theory. I do not think that they do.

- Darwin himself had considerable doubts; his book contains a chapter called 'Difficulties on Theory'. Of particular interest to the physicist are his remarks about the eye: 'To suppose that the eye, with its inimitable contrivances for adjusting the focus to different distances, for admitting different amounts of light, and for the correction of spherical and chromatic aberration, could have been formed by natural selection, seems, I freely confess, absurd in the highest possible degree.' Nevertheless, he goes on to suggest how an eye *could* have developed from a simple light-sensitive organ, but there is no evidence that it *did* develop in that way.

- If living matter is not, then, caused by the interplay of atoms, natural forces and radiation, how has it come into being? . . . I think, however, that we must go further than this and admit that the only acceptable explanation is *creation*. I know that is anathema to physicists, as indeed it is to me, but we must not reject a theory that we do not like if the experimental evidence supports it.[37]

Lipson is convinced that the scientific evidence points to creation and not evolution. In the next chapter we shall consider how God's historical claim to have created the world strengthens the creationist worldview and weakens the evolutionist worldview.

NOTES

1. Introduction, *The Origin of Species* (London: J. M. Dent & Sons, 1971), pp.x-xi.
2. "How Evolution Became a Religion: Creationists Correct?" *National Post*, 13 May 2000, p. B1.
3. Science and Religion: From Conflict to Conversation (New York: Paulist Press, 1995), p. 56.
4. Ibid, pp. 33-34.
5. Ibid, p. 56.
6. "The Origin of the Universe in Science and Religion," *Cosmos, Bios, Theos*, ed. Henry Margenau and Roy Abraham Varghese (La Salle, IL: Open Court, 1992), p. 254.
7. *From Evolution to Creation: A Personal Testimony* (Waterloo: Answers in Genesis, 2000), pp. 12-13.
8. "Billions and Billions of Demons," *The New York Review*, 9 January 1997, p. 31.
9. See Richard Dawkins, *The God Delusion* (Boston: Houghton Mifflin, 2006), pp. 69-73; also Joseph.Brean, "Evolutionary Biologist Professes Belief in Aliens," *National Post*, 21 June 2007, p. A10.
10. "Evolutionary History and Population Biology," *Nature* 214 (1967), p. 352.
11. Science and Creationism: A View from the National Academy of Sciences, 2nd ed. (Washington, D.C.: National Academy Press, 1999), p. 2.
12. *The Phenomenon of Man* (New York: Perennial, 2002), pp. 218-19.
13. *Origins: Linking Science and Scripture* (Hagerstown, MD: Herald Publishing, 1998), p. 17.
14. Phil Gibbs and Hiroshi Sugihara, "What Is Occam's Razor?" http://math.ucr.edu/home/baez/physics/General/occam.html. Retrieved 26 December 2008.
15. *Introducing Evolution* (Triplow: Incon Books, 2001), pp. 45-46.
16. "Social Responsibility and the Scientist," *Perspectives in Biology and Medicine* 14 (1971), p. 353.
17. What Mad Pursuit: A Personal View of Scientific Discovery (New York: Basic Books, 1990), p. 138.

18. *Defeating Darwinism by Opening Minds* (Downers Grove, IL: Intervarsity Press, 1997), p. 114.

19. *Darwin and Intelligent Design* (Minneapolis: Fortress Press, 2006), p. 73.

20. *Creation: Facts of Life* (Green Forest, AR: Master Books 2006), p. 52.

21. Evolution and the Myth of Creationism: A Basic Guide to the Facts in the Evolution Debate (Stanford: Stanford University Press, 1990), pp. 118-19.

22. Defeating Darwinism by Opening Minds, p. 63

23. Ibid, p. 126.

24. *The Complete Idiot's Guide to Evolution* (Indianapolis: Alpha Books, 2002), p. 7.

25. "Social Responsibility and the Scientist," p. 367.

26. Creation: Facts of Life, pp. 138-39.

27. "Social Responsibility and the Scientist," pp. 368-69.

28. "The Second Evolution: The Unified Teem Theory of Evolution, Perception, Emotions, Behaviour and Inheritance," http://thesecondevolution.com/Introduction.pdf. Retrieved on 3 April 2011.

29. *Evolution: A Theory in Crisis* (Bethesda, MD: Adler and Adler, 1986), p. 358.

30. *The American Spectator* 27.7 (July 1994), p. 16. http://www.geocities.com/truedino/newdogma.htm. Retrieved 1 September 2008.

31. A. E. Wilder-Smith, *The Scientific Alternative to Neo-Darwinian Evolutionary Theory* (Costa Mesa, CA: The Word For Today, 1987), p. xii.

32. Why Darwin Matters: The Case Against Intelligent Design (New York: Henry Holt, 2006), p. 166.

33. Ibid, p. 167.

34. "Evolution, Creation, and Biology Teaching," *Evolution Versus Creationism: The Public Education Controversy*, ed. J. P. Zetterberg (Phoenix: Oryx Press, 1983), p. 91.

35. The Phenomenon of Man, pp. 218-19.

36. Ibid.

37. "A Physicist Looks at Evolution," *Physics Bulletin* 31 (May 1980), p. 138.

CHAPTER 5

EXALTING THE CREATIONIST WORLDVIEW

This is what the LORD, the Creator and Holy One of Israel, says: . . . I am the one who made the earth and created people to live on it. With my hands I stretched out the heavens.

—Isaiah 45:11-12 NLT

An animal—particularly the human animal—is a beautiful example of a carefully contrived and subtly engineered design. The word 'design' comes naturally even in evolutionist books. The Designer must know infinitely more science than we shall ever know.[1]

—H. S. Lipson

By analogy we are supernatural beings to earthworms, which are blind to our world. The human being can boast of executing tasks such as laying the foundation, putting up a roof, and furnishing an abode building in a matter of a week; to earthworms, these claims would be preposterous. God, who is supernatural to human beings, in turn, can boast of executing tasks such as laying the foundation of the earth, stretching the sky with its heavenly bodies (Isaiah 44:24, 45:12, 48:13; Jeremiah 10:12) and furnishing the universe and creating various life forms on it within six days; to human beings these claims are preposterous.[2]

—Michael Ebifegha

CAN THE DESIGNED CHOOSE ITS DESIGNER?

When a fetus gets to choose its parents, perhaps then we can believe that scientists are in a position to choose who or what designed us. Until then we cannot treat science as the absolute guide to truth; we must open our minds to scriptural revelations in the course of history.

According to the biblical account, God organized the heavens and the earth to accommodate life in six days (Gen. 1; Exod. 20:11) through a combination of supernatural power, wisdom, and understanding (Jer. 10:11-12). The Scriptures declare that the earth hangs on *nothing* (Job 26:7) and is intentionally designed to be inhabited (Isa. 45:18). Humankind was created in the image of God with the mandate to subdue the earth and rule over everything (Gen. 1:26-28). According to the Scriptures some of our earliest ancestors lived over 900 years, indicating that the world began in relatively perfect condition. Unfortunately, following human intransigence culminating in sin against God, the earth and its inhabitants experienced corruption and have been deteriorating ever since. As a result, we inherit a universe that is plagued with problems such as tsunamis, disease, and earthquakes. Evidently these events in nature lead secular scientists to believe that the universe is the product of chance, accident, and blind forces.

In its 1995 statement on teaching evolution, for example, the American National Association of Biology Teachers (NABT) declared: "The diversity of life on earth is the outcome of evolution: an unsupervised, impersonal, unpredictable, and natural process of temporal descent with genetic modification that is affected by natural selection, chance, historical contingencies, and changing environment."[3] To avoid any religious implications, the terms "unsupervised," "impersonal," and "chance" were omitted in the 2004 version: "The diversity of life on earth is the outcome of biological evolution—an unpredictable and natural process of descent with modification that is affected by natural selection, mutation, genetic drift, migration, and other natural biological and geological forces."[4] NABT is careful about its choice of words. Brian Killian argues: "A definition of evolution that uses these adjectives ("unsupervised" and "impersonal," similar to "unguided" and "unplanned") is not, in fact, a biological definition. It is more like a mission statement for atheism masquerading as biology."[5]

Divergent views about the origin of life have split the scientific community into two camps—creationists and evolutionists. Both groups agree that the universe is designed, but they disagree on the nature of the designer. Creationism and evolutionism share a common weakness: their advocates were not present at the time of the alleged events to justify their choice. Creationism, however, has a unique strength in positing a veridical claimant. This uniqueness favours the creationist stance and debunks the evolutionist worldview.

The origin and diversity of life on earth, like the mystery of love, are a phenomenon that science cannot illuminate as successfully as other aspects of biology. Francisco Ayala points out in *Darwin and Intelligent Design* that "science is a powerful and successful way of acquiring knowledge about the universe, but it is not the only way: other valid ways of acquiring knowledge about the universe include imaginative literature and other forms of art, common sense, philosophy, and religion."[6] Because the events pertaining to "origin" are not repeatable, science alone cannot provide the correct answer.

When the choice is between two designing agencies, the conclusions are philosophical; hence, scientists are unlikely to be unanimous in their choice. Science, therefore, is not a reliable and apodictic guide. According to Albert Einstein, "Science without religion is lame[;] religion without science is blind."[7] Therefore, input from both science and religion is required to figure out the truth. This chapter focuses on the religious framework.

When the choice is between an intelligent designer and a non-intelligent instrumentality, the problem comes down to common-sense knowledge. One important difference between a designer and an instrumentality is the ability to communicate. The ability to communicate information in verbal, coded, or printed form is one of the universal measures of intelligence. A credible and responsible designer would publicly claim credit in words and/or print.

If we are made in the image of a creator, logic indicates that our creator is superior to us and must be personal to claim us as moral beings. Unlike robots, we are conscious and therefore seek to understand our ontological origins. In this effort, however, we must admit our limitations and not override them with bogus assumptions about life's arising from non-life (contradicting empirical evidence), programmed information without a programmer (antithetical to common sense), blueprints emerging as the architect (as though our genes created us), mindless processes generating spirituality (science does not deal with the immaterial world), and so on. If these same claims were part of Scripture, they would have been deemed "miracles." And, according to Richard Dawkins, "Any belief in miracles is flat contradictory not just to the facts of science but to the spirit of science."[8]

In order to silence creationists who proclaim a creator, John Rennie, the editor of *Scientific American*, writes in "15 Answers to Creationist Nonsense": "If superintelligent aliens appeared and claimed credit for creating life on earth (or even particular species), the purely evolutionary explanation would be cast in doubt. But no one has yet produced such evidence."[9] Rennie's demand for evidence is the primary reason for my previous book, *The Death of Evolution: God's Creation Patent and Seal*. A claim to have created the universe cannot be brushed aside as mere religious superstition. If scientists are concerned enough to rebut the Creation account in the first and second chapters of Genesis, they also need to address God's public claim to have created the universe in the nineteenth and twentieth chapters of Exodus.

The Judaeo-Christian Scriptures are the expression of God's claims, and we cannot ignore them. If one explores them with an open mind, one can establish the truth of our origin. As in every other history of world events, people, most of whom witnessed the reported events, wrote the Bible. Most historical accounts are written after the fact; temporal delays do not invalidate them. The historical accounts in the Bible are authentic records, *not* fanciful myths. The following epilogue of a book titled *The Miracles of Exodus* by Colin J. Humphreys, a physicist at Cambridge University, is relevant here:

> At the start of this book, I asked a number of important questions. Let's see if we can now answer them. First, is the Old Testament account of the Exodus from Egypt a coherent and consistent account (many scholars believe it isn't)? My answer to this is a resounding yes. Second, is the Exodus account in the Bible factually accurate? When I started my research on this book in 1995, I really wondered just how accurate the biblical text was. I was well aware that most scholars believe it is riddled with errors and inconsistencies. I've subjected the biblical text to a real grilling in this book, and I can only stand back in amazement at its accuracy and consistency, down to points of tiny detail . . . . Finally, I asked in the first chapter, Is there any evidence of a "guiding hand" in the events of the Exodus? What I've found is that the Exodus story describes a series of natural events like earthquakes, volcanoes, hail and strong winds occurring time after time at precisely the right moment for the deliverance of Moses and the Israelites. Any one of these events occurring at the right time could be ascribed to lucky chance. When a whole sequence of events happens at just the right moment, then it is either incredibly lucky chance or else there is a God who works in, with, and through natural events to guide the affairs and the destinies of individual and of nations. Which belief is correct: Chance or God? . . . Please think long and hard because if the arguments I've given are correct, then this book has rewritten our understanding of a major event in world history: the Exodus from Egypt.[10]

The prophetic accounts in Scriptures also are not fantasies or myths. The enslaved Children of Israel in Egypt, their exile throughout the world, and their subsequent return as a nation are fulfilled prophesies. If, according to atheist Victor J. Stenger, "prophecies were inserted after the fact,"[11] the establishment of the modern state of Israel in 1948 proves how biased his views are about God. Israel remains a shining star in the modern world as it was in the ancient one. Thomas Cahill, stressing this fact in his book *The Gift of the Jews*, stipulates:

> The story of Jewish identity across the millennia against impossible odds is a unique miracle of cultural survival. Where are the Sumerians, the Babylonians, the Assyrians today? And though we recognize Egypt and Greece as still belonging to our world, the cultures and ethnic stocks of those countries have little continuity with their ancient namesakes. But however miraculous Jewish survival may be, the greater miracle is surely that the Jews developed a whole new way of experiencing reality, the only alternative to all ancient worldviews and all ancient religions. If one is ever to find the finger of God in human affairs, one must find it here . . . . All religions are cyclical, mythical, and without reference to history as we have come to understand it—all religions *except* the Judeo-Christian stream in which Western consciousness took life.[12]

Here Cahill affirms historian Nicholas Berdyaev's conclusion: "The philosophy of man's terrestrial destiny may be said to begin with the philosophy of history and destiny of the Jewish people. Here lies the axis of world history.[13]

Quite apart from the above reasons, Dawkins in *The God Delusion* makes reference to a number of passages in the Judaeo-Christian Scriptures, especially from Exodus. The present book as a rebuttal will address some of the passages Dawkins uses to support his thesis that "there almost certainly is no God."[14]

UNDERSTANDING THE JUDAEO-CHRISTIAN GOD

We should begin with an introduction to God because this book is about the "God of Creation," not the "God of Evolution" that some scholars proclaim.[15] The Judeo-Christian God, the One that organized the universe in six days, explicitly declared His sovereignty: "I am the First and I am the Last, the Alpha and the Omega, the Beginning and the End" (Isa. 41:4, 44:6, 48:12; Rev. 1:8, 21:6, 22:13).

This description of God relates to our universe that also has a beginning and an end. As "the Beginning and the End," God is the *uncaused* cause behind every created thing, visible or invisible. In so defining Himself, God refutes all scientific theories of evolution that grant undue credit to blind natural processes. God is the author of light and therefore of darkness as a special creation. God exclusively created *ex nihilo* through the power of words, and since words are something it would imply that "nothing" is also a special creation. LeeAundra Temescu discuses the mysteries of this concept:

> There is vastly more nothing than something. Roughly 74 percent of the universe is "nothing," or what physicists call dark energy; 22 percent is dark matter, particles we cannot see. Only 4 percent is baryonic matter, the stuff we call something.[16]

And to affirm that "nothing" is part of God's design, the Scriptures intimate that God hangs the earth on nothing (Job 26:7). So, when scientists suggest that the universe evolved from "nothing," they are in essence saying that the universe is a product of God's creation.

Given God's self-definition as "the Beginning and the End," Richard Dawkins' denigrating question "Who designed the designer?"[17] is fatuous. The same is true of his comment, "A designer God cannot be used to explain organized complexity because any God capable of designing anything would have to be complex enough to demand the same kind of explanation in his own right."[18] Furthermore, as "nothing" is another aspect of God's invention, God also debunks Peter Atkins' contention that "The universe can emerge out of *nothing*, without intervention. By chance."[19] Stenger argues in *God: The Failed Hypothesis* that "Only by the constant action of an agent outside the universe, such as God, could a state of nothingness be maintained."[20] Stenger thus confirms rather than denies God's existence. Atheist Taner Edis contends that "In all likelihood, the universe is uncaused."[21] The universe is uncaused in the sense that God fills the universe, but according to science we see only about 4% of the universe as "something." Einstein is right in saying that God is an illimitable Spirit! The Scriptures read: "Am I only a God nearby, and not a God far away? Can anyone hide in secret places so that I cannot see him? Do not I fill heaven and earth?" (Jer. 23:23-24 NIV).

It seems that modern atheists let their personal beliefs overrule their objective analysis of any data, whether scientific or religious. In another determined effort to deny God's existence, for example, Stenger writes:

> A God who provides humans with important knowledge that they cannot obtain by material means should have produced testable evidence for his existence by now. He has not. The evidence points to the opposite. We can say with some confidence that such a God does not exist.[22]

By a similar line of reasoning, human beings have created robots and other things that are without testable evidence for human existence. Can these objects say with some confidence that human beings do not exist? The late Nobel laureate Christian B. Anfinsen, presumably vexed by such pointless arguments, lashed out:

> I think only an idiot can be an atheist. We must admit that there exists an incomprehensible power or force with limitless foresight and knowledge that started the whole universe going in the first place.[23]

I do not think that atheists are idiots, but I believe they lack spiritual insight, the kind that Einstein consistently linked with science. Atheists are looking at the evidence through the wrong lens. They need to switch to a spiritual lens in order to embrace the unequivocal evidence that points to a God who endowed us with spiritual values. For over fifty years Antony Flew, a world-renowned atheist, viewed scientific evidence through the lens of materialism. With the appropriate lens, however, Flew found that the same evidence points to God's existence. God is Spirit and should be sought accordingly.

I will use an empirical evidence to explain the difference between a spiritual-minded and material or secular-minded scientist. Dr. Stephen Grocott, Fellow of the Royal Australian Chemical Institute, describes the complexity of the simplest organism as follows:

> The complexity of the simplest imaginable living organism is mind-boggling. You need to have the cell wall, the energy system, a system of self-repair, a reproduction system, and means for taking in 'food' and expelling 'waste', a means for interpreting the complex genetic code and replicating it, etc., etc. The combined telecommunication systems of the world are far less complex, and yet no one believes they arose by chance.[24]

The spiritually inspired scientist, like Einstein, would acknowledge this statement as evidence of illimitable intelligence. To reach this conclusion, one has to rely only on common sense and simple analogies in the real world. The materialist scientist, like Dawkins, would explain this as evidence of chance and cumulative selection. Common sense indicates that such an explanation is highly unlikely, but to an atheist materialism is the governing creed. Accordingly, as most leading scientists today are atheists, the modern scientific community endorses Dawkins' views over Einstein's in order to privilege materialism. Philosopher Deepak Chopra remarks:

> [Einstein's] God was too impersonal for the religionists; his physics was too idealistic for the scientists (idealistic

in the sense that Einstein never abandoned his belief in a non-random creation). What's so heartening about him today is that Einstein never adopted the arrogant small-mindedness of contemporary atheists like Richard Dawkins and Christopher Hitchens, neither of whom evinces the slightest awareness of the quantum revolution that occurred a century ago. Without a shadow of arrogance Einstein wrote, "What separates me from most so-called atheists is a feeling of utter humility toward the unattainable secrets of the harmony of the cosmos." I remain amazed that his beliefs are dismissed while those of much lesser minds earn general acceptance. Quite rightly, Einstein thought that atheists are slaves to religious tradition they hate and hold such a grudge against traditional religion that "they cannot hear the music of the spheres."[25]

Ernst Boris Chain expressed a similar concern:

These classical evolutionary theories are a gross oversimplification of an immensely complex and intricate mass of facts, and it amazes me that they were swallowed so uncritically and readily, and for such a long time, by so many scientists without a murmur of protest.[26]

Modern atheists are bent on using science to dismiss the notion of a living God. The irony is that their interpretation of scientific evidence trivializes the scientific enterprise.

In assigning to Moses the task of freeing the Israelites from slavery in Egypt, God announced Himself as the "I AM" (Exod. 3:14). The declarative simply means always present or timeless, the limitless, self-existent One "without beginning or end." David Berlinski in *The Devil's Delusion*, referring to this unique claim, writes:

The five simple words of the declaration in Exodus—"I am that I am"—suggest that God's existence is necessary.

> Being what He is, God could not fail to be who He is,
> and being who He is, God could not fail to *be* . . . .
> Thus there is one thing whose existence is necessary,
> and if necessary, by the very same argument, eternal.
> Since it is eternal, it has no cause. Questions about *its*
> origins are pointless.[27]

Berlinski points to the consistency and perpetuity of God. Science has not and cannot address these attributes of God, and this defeats the argument that science can deal with every aspect of existence. Dawkins' claim that God's existence is a scientific question is thus baseless.[28]

God dwells within and beyond the universe. This limitless God, in any intervention such as appearing to individuals through dreams and visions or talking to a crowd of people, assumes any form or symbol of presence depending on the occasion. Einstein recognized God as an illimitable Spirit but failed to realize that this limitless God can assume any manifestation. Our images of God as He or She, Father or Mother, are simply expressions of the living Supernatural Being in a supreme position of authority. Chopra's view on this point is noteworthy:

> God, if he exists, is universal, existing at all times and
> places, pervading creation both inside the envelope of
> space-time and outside it. To use a word like "He" has
> no validity, in fact; we are forced into it by how language
> works. A better word would be "the All," which in Sanskrit
> is Brahman and Allah in Islam. Not every language is
> stuck with He or She.[29]

Chopra admonishes his readers to forget anthropomorphic images of God: "Forget the image of God sitting on a throne, forget Genesis, forget the straw man of a Creator who isn't as smart as a smart human being."[30] We, of course, cannot forget Genesis. The important point is that, if humans are made in the image of God, certain attributes such as love, judgment, communication, and moral values should reflect the image of

God. God can be an illimitable Spirit and still possess unique personality.

Many scientists and lay persons consider the Genesis account of Creation an ancient myth to justify their evolutionist worldview. To steer clear of arguments about the Genesis report, our focus here is on the Exodus cosmological account in which God personally claims credit for having created the universe before an assembly of ancient Israelites. In this account God specifies the duration of six days for preparing the earth for habitation and sets His seal on the seventh day. These claims are presented verbally in a set of ten universal commandments (Decalogue), which are not myths. In addition, God engraved these commandments on tablets of stones (Exod. 31:18). Thus, while sceptics may brush aside the Genesis report as an ancient myth, the Exodus account is consistent with the modern protocol personal claims.

THE CREATION SABBATH COMMANDMENT

God issued the Ten Commandments on Mount Sinai. The fourth Commandment, "The Creation Sabbath," presents God's claim to have created the universe:

> Remember the Sabbath day by keeping it holy. Six days you shall labour and do all your work, but the seventh day is a Sabbath to the LORD your God. On it you shall not do any work, neither you, nor your son or daughter, nor your manservant or maidservant, nor your animals, nor the alien within your gates. For in six days the LORD made the heavens and the earth, the sea, and all that is in them, but rested on the seventh day. Therefore the LORD blessed the Sabbath day and made it holy. (Exod. 20:8-11 NIV)

God stopped working ("rested") on the seventh day to attest that the supernatural formation of the universe was completed. The primordial universe was endowed with appropriate laws for continuous structural expansion and the reproduction of life forms. For reasons outside the scope of science, God chose six days to create the universe and living organisms. The seventh day is God's eternal seal on Creation. Its "eternal" nature is indicated by the absence of an "evening" and a "morning" that characterized each of the six days of creative work described in Genesis 1. For human beings God ordained the seventh day as a Sabbath, which completes the weekly cycle that serves as a reminder of God's creative labour. The Sabbath connects the beginning of the first day and the end of the sixth day to complete our seven-day weekly cycle.

The rotation of the earth on its axis determines the day's length (24 hours). The moon's orbit around the earth determines the month's length (about 708 hours). The earth's revolution around the sun determines the year's length (about 8760 hours). However, unlike the day, month, and year, the seven-day week (168 hours) has no astronomical basis. In other words, the day, month, and year are derived from the material world, but the week has no material basis. Since God revealed the seventh day and imposed its observation as a moral commandment, it follows that the seven-day weekly cycle has a purely divine origin and is the temporal link between the material (physical) and non-material (spiritual) world. Its divine origin has proven, from time to time, anathema to both politicians and scientists.

In the political realm, in order to undermine the supernatural origin of our weekly cycle, secular governments have tried to alter the weekly period from seven days to five, six, eight, or ten days. Eventually they reverted to the divinely assigned seven-day cycle. Susan Perry and Jim Dawson thus report:

> After the French Revolution at the end of the eighteen century, the new rulers of France tried to abolish the seven-day week, which they thought was rooted in religious superstition, in favour of a more "rational" ten-day week. But the experiment failed. People continued to take a day of rest every seven rather than every ten days. More than one hundred years later, revolutionary

leaders in the Soviet Union made a similar attempt to change the week—first to five days, then to six. Again, people resisted, and the seven-day week eventually was re-established.[31]

In the scientific realm the discovery of chronobiology (biological rhythms) shows that (1) key biological functions in humans such as heartbeat, variations in blood pressure, and response to infection exhibit a seven-day weekly rhythm; and (2) plants, insects, and animals follow seven-day biorhythmic cycles. In this regard Perry and Dawson write:

> Weekly rhythms—known in chronobiology as *circaseptan rhythms*—are one of the most puzzling and fascinating findings of chronobiology. Daily and seasonal cycles have an obvious link to the sun, and monthly cycles appear to be connected to the moon. But what is there in nature that would have caused weekly rhythms to evolve? At first glance, it might seem that weekly rhythms developed in response to the seven-day week imposed by human culture thousands of years ago. However, this theory doesn't hold once you realize that plants, insects, and animals other than humans also have weekly cycles . . . . Biology, therefore, not culture, is probably at the source of our seven-day week. It certainly is a rhythm deeply ingrained within us . . . . Scientists now theorize that our social week may actually be a *Zeitgeber* (time-giver), helping to reset our weekly biological rhythms—just as our daily social routines help reset our daily rhythms. That may explain why some of us who are used to relaxing on Saturday and Sunday feel so disoriented when we must work through the weekend. We have disrupted our rhythms.[32]

With God as the originator of the seven-day weekly cycle, these discoveries corroborate a creationist as opposed to an evolutionist worldview. Dwight K. Nelson agrees:

Science cannot explain the seven-day week except by appealing to history. And history declares that the most consistent accounting for our . . . seven-day week is found in the ancient Hebrew recording of creation. Thus, the seven-day week and the seventh-day Sabbath are a perpetual testimony to the veracity and historicity of the creation accounts in Genesis 1 and 2. Not only do we have both of them affirmed here in the Creation accounts of Genesis, but we must also reckon with the indisputable linkage to these accounts in the very heart and soul of the divinely composed Ten Commandments.[33]

Exodus provides an abstract of God's record of Creation, and Genesis gives a concise account of the methodology and results. Although science is unable to explain the seven-day weekly cycle, it has uncovered its biological relevance. Since the Creation Sabbath Commandment stipulates that domestic animals must also rest on the Sabbath day, scientific findings are consistent with God's declaration. Science thus lends support to the historical evidence of God's existence!

THE CREATION SABBATH LAW

The Creation Sabbath Commandment consists of instructions elaborated into a civil law. It reads:

Observe the Sabbath, because it is holy to you. Any one who desecrates it must be put to death; whoever does any work on that day must be cut off from his people. For six days, work is to be done, but the seventh day is a Sabbath of rest, holy to the LORD. Whoever does any work on the Sabbath day must be put to death. The Israelites are to observe the Sabbath, celebrating it for the generations to come as a lasting covenant. It will be a sign between me and the Israelites forever, for in six days the LORD made the heavens and the earth, and on the seventh day abstained from work and rested. (Exod. 31:14-17 NIV)

For forty years the Israelites sojourned in the desert, subsisting on the special manna diet God supplied daily. To identify which day was the Sabbath, God withheld the daily ration on the Sabbath but doubled it on the preceding day. The double portion of food was to ensure that, on the Creation Sabbath, the Israelites were relieved from the labour of food collection and preparation.

The following points clarify why this historical evidence cannot be ignored or dismissed as a myth:

♦ *The evidence is credible.*

Before an assembly of ancient Israelites, God verbally claimed credit for having created the universe. It was presented as God's Creation Sabbath Commandment. Eyewitnesses passed on this experience to subsequent generations (Deut. 4:9-16, 32-36; 5:4, 22-30), a process that continues to this day. The provision of the sacred books ensures the consistency of message—i.e., the Pentateuch and secular books on Jewish history.

♦ *The evidence is secular.*

God's Creation Sabbath Commandment is consistent with our world's standard for patenting. Our tradition is that ownership of a property or exclusive rights to an invention are documented in public. Almighty God publicized a summary or abstract of having created the universe in the Creation Sabbath Commandment, but the details of sequential events are reported in Genesis.

♦ *The evidence is conclusive.*

God's claim to have created the universe is not only historical but also binding on all people. The Bible, the primary documentation of this event, does not limit its scope and implications to the religious domain; it is equally a civil or legal matter since the claim is universal. This historical event of God's unique intervention is proof that the Judaeo-Christian God is personal, living, and active.

DAWKINS' FAILURE TO DISPOSE OF THE ARGUMENT FOR GOD'S EXISTENCE

The Creation Sabbath is the cornerstone of God's Ten Commandments because it is the only commandment that simultaneously addresses our obligation to God as the Creator and to others as our neighbours. In its comments on this historical event, the Jewish Study Bible asserts:

> The momentous encounter with God at Sinai is, for Judaism, the defining moment in Jewish history, the moment when God came down to earth and spoke to all Jews, present and future, giving them His rules for life, which they accepted enthusiastically.[34]

In his discussion of the same event Rabbi Aryeh Kaplan writes:

> It is the Exodus that makes Judaism unique. God revealed Himself to an entire people, and literally changed the course of nature for a forty-year period. This was an event unique in the history of mankind . . . . There are other religions in the world, but none of them can match the powerful beginnings of Judaism. The others all began with a single individual, who claimed to have spoken to God or arrived at Truth. This individual gradually spread his teachings, forming the basis of a new religion. Virtually every great religion follows this pattern. Judaism is unique in that God spoke to an entire people, three million people at the same time, who saw with their own eyes and heard with their own ears. That one historic, traumatic experience is the solid bedrock of Jewish faith. The Exodus not only made us uniquely aware of

God, but it also showed Him profoundly involved in the affairs of man.[35]

Kaplan continues:

God was telling us that He is involved in the affairs of man and has a profound interest in what we do. God Himself gave the Exodus as an example. It was here that the entire Jewish people experienced God. To them, God was no mere abstraction. They saw His deeds to such an extent that they were actually able to point and say, "This is my God."[36]

Although Exodus provides an historical account of God's dealing in person with the human race, Richard Dawkins fails to mention this point in *The God Delusion*, preferring instead to fulminate against essentially peripheral issues.

For example, Dawkins narrates the incident in which a Sabbath profaner was stoned to death at God's command. His interest in this matter is driven by an effort to justify his view that the Judaeo-Christian God is a bully. His emotionally surcharged rhetoric gives the game away: "What makes my jaw drop is that people today should base their lives on such an appalling role model as Yahweh—and, even worse, that they should bossily try to force the same evil monster (whether fact or fiction) on the rest of us."[37] In fairness one cannot say for certain that Dawkins is aware of the history surrounding God's Creation Sabbath, but his bias is unmistakable.

Dawkins nonetheless has avoided any discussion of the events that led to God's verdict on those who broke the Creation Sabbath Covenant, and he has done so presumably for three reasons. First, he does not believe in God's six days of Creation. Capital punishment implies capital offence, and in this case the offence pertains to disregard of God's decree about breaking the Creation Sabbath Law. Second, as a die-hard atheist Dawkins does not believe in the existence of God. Third, Dawkins wants to dissuade his larger audience from believing in God.

To his credit Dawkins admits that there are problems science cannot solve: "Perhaps there are some genuinely profound and meaningful questions that are forever beyond the reach of science. [. . .]. But if science cannot answer some ultimate question, what makes anybody think that religion can?"[38] Dawkins is right about the inability of science to answer certain questions, but he is wrong in concluding that religion cannot answer them. The origin of species is one of the questions that science cannot answer.

God, like the human mind, is immaterial. The human mind exists but we cannot see, touch, smell, hear or taste it, hence, it is outside science's purview. No scientist operating within the limits of science can claim that God is a delusion. A scientific theory, however, may contain scientific facts, but could still be wrong by imposing the wrong set of assumptions and issuing the incorrect interpretation of the evidence. Thus, while the scientific evidence cannot be used to show that God or the creationist worldview is a delusion, it can be used to disqualify the evolutionist worldview that is based exclusively on materialism.

In sum, "While science has nothing of value to say on the great and aching questions of life, death, love, and meaning, what the religious traditions of mankind *have* said forms a coherent body of thought."[39] We must appreciate the fact that science is limited and that it is not the only way to arrive at truth.

NOTES

1. "A Physicist Looks at Evolution," *Physics Bulletin* 31 (May 1980), p. 138.
2. The Death of Evolution: God's Creation Patent and Seal (Longwood, FL: Xulon Press, 2007), pp. 103-04.
3. Quoted in Phillip E. Johnson, *Defeating Darwinism by Opening Minds* (Downers Grove, IL: Intervarsity Press, 1997), p. 15. I could not get hold of the 1995 version, instead I received that of 2004 by e—mail NABT on July 31, 2007.
4. Email communication of 31 July 2007 from the American National Association of Biology Teachers.
5. "The Other Creation Story," http://www.catholiceducation.org/articles/science/sc0065.html. Retrieved March 19 2011.
6. *Darwin and Intelligent Design* (Minneapolis: Fortress Press, 2006), p. 90.
7. *The New Quotable Einstein*, ed. Alice Calaprice (Princeton: Princeton University Press, 2005), p. 203.
8. Quoted in D. V. Biema, "God vs. Science," *Time*, 13 November 2006, p. 36.
9. "15 Answers to Creationist Nonsense," *Scientific American* 287.1 (2002), p. 80.
10. *The Miracles of Exodus* (New York: HarperCollins, 2003), pp. 339-40. Professor Colin J. Humphreys, is a world-renowned Cambridge University physicist, president of the Institute of Materials Science and Goldsmiths' Professor of Materials Science, and head of the Rolls Royce University Technology Centre. He is also an expert in chemistry, astronomy, and geology. He has been examining the Bible in the light of science for over twenty years.
11. *God: The Failed Hypothesis. How Science Shows That God Does Not Exist* (New York: Prometheus Books, 2007), p. 182.
12. *The Gift of the Jews: How a Tribe of Desert Nomads Changed the Way Everyone Thinks and Feels* (New York: Anchor Books, 1999), pp. 246, 248.
13. *The Meaning of History*, trans. G. Reavey (Glasgow: The University Press 1936), pp. 86-87.

14. *The God Delusion* (Boston: Houghton Mifflin, 2006), pp. 73, 111.

15. Denis Edwards, *The God of Evolution* (New York: Paulist Press, 1999).

16. "20 Things You Didn't Know About Nothing," *Discover*, June 2007, p. 88.

17. *The God Delusion*, p. 121.

18. Ibid, p. 109.

19. *Creation Revisited* (London: Penguin, 1994), p. 143.

20. *God: The Failed Hypothesis*, p. 133.

21. *The Ghost in the Universe: God in the Light of Modern* Science (New York: Prometheus Books, 2002), p. 97.

22. *God: The Failed Hypothesis*, p. 133.

23. "There Exists an Incomprehensible Power with Limitless Foresight and Knowledge," *Cosmos, Bios, Theos*, ed. Henry Margenau and Roy Abraham Varghese (La Salle, IL: Open Court, 1992), p. 138.

24. "Science and Origins," *In Six Days: Why 50 Scientists Choose to Believe in Creation*, ed. John F Ashton (Sydney: New Holland Publishers, 1999), p. 136.

25. "Einstein's God, or The Hopes for a Secular Spirituality (Part 4)," intentBlog, 7 September 2007. www.deepakchopra.com.

26. "Social Responsibility and the Scientist," *Perspectives in Biology and Medicine in Modern Western Society* 14 (1971), pp. 367-68.

27. The Devil's Delusion: Atheism and Its Scientific Pretensions (New York: Crown Forum, 2008), pp. 84-85.

28. See "God vs. Science," p. 35.

29. "The God Delusion? (Part 2)," intentBlog, 17 November 2006. www.deepakchopra.com.

30. Ibid.

31. *The Secrets Our Body Clocks Reveal* (New York: Rawson Associates, 1988), p. 21.

32. Ibid, pp. 20-21.

33. *Built to Last: Creation and Evolution* (Nampa, Idaho: Pacific Press, 1998), p. 127.

34. *The Jewish Study Bible* (New York: Oxford University Press, 1999), p.106.
35. "Why the Sabbath-Sabbath Day of Eternity?" http://www. ou.org/publications/ kaplan/shabbat/why.htm.
36. Ibid.
37. The God Delusion, p. 248.
38. Ibid, p. 59.
39. David Berlinski, *The Devil's Delusion*, p. xiv.

CHAPTER 6

THE NATURAL SELECTION DELUSION

To postulate . . . that the development and survival of the fittest is *entirely* a consequence of chance mutations, or even that nature carries out experiments by trial and error through mutations in order to create living systems better fitted to survive, seems to me a hypothesis based on no evidence and irreconcilable with the facts.[1]

—Ernst Boris Chain

Natural selection can serve in nature to eliminate aberrant types, but not to produce new complex structures that would not have survival value until all necessary parts have evolved to form a functional system.[2]

—Ariel A. Roth

Today our duty is to destroy the myth of evolution, considered as a simple, understood, and explained phenomenon which keeps rapidly unfolding before us. Biologists must be encouraged to think about the weaknesses of the interpretations and extrapolations that theoreticians put forward or lay down as established truths.[3]

—Pierre-P. Grassé

I want to know how God created this world. I'm not interested in this or that phenomenon, in the spectrum of this or that element. I want to know His thoughts[;] the rest are details.[4]

—Albert Einstein

Evolutionists contend that the universe happened by accident; that it was established several billion years ago following a gigantic explosion of concentrated mass and energy; and that complex life evolved over millions of years when lifeless debris was resurrected by chance and fashioned in the workshop of random natural processes by genetic mutation and natural selection. Concerning this theory Theodosius Dobzhansky writes:

> Evolution is not predestined to promote always the good and the beautiful. Nevertheless, evolution is a process which has produced life from non-life, which has brought forth man from an animal, and which may conceivably continue doing remarkable things in the future . . . . Evolution comprises all the stages of development of the universe: the cosmic, biological, and human or cultural developments . . . . Life is a product of the evolution of inorganic nature, and man is a product of the evolution of life . . . . Mankind's distinctive attributes and capacities arose in evolution under the control of natural selection. Natural selection makes the evolutionary changes usually adaptive in the environments in which the species lives.[5]

There is no evidence, however, to suggest that evolution can produce life from non-life. There is also no empirical support to the assertion that natural selection controls humanity's distinctive attributes.

According to NASIM, explanations in science must be based on naturally occurring phenomena, but life arising from non-life (abiogenesis) is not a naturally occurring phenomenon, so any explanation based on that premise is unscientific. NASIM also contends that "natural causes are, in principle, reproducible and, therefore, can be checked independently by others."[6] Bacteria-to-human evolution is not reproducible and cannot be verified in any laboratory; hence, it cannot be due to natural causes. Evolutionism, like creationism, is thus unscientific.

COURTING NATURAL SELECTION TO SNEAK IN ATHEISM

Science is simply a branch of knowledge that human beings have developed to explain the composition and function of matter. Its success within such parameters, however, cannot be stretched into other domains. For instance, since the brain is composed of matter, biological science can study its composition and functions. The *mind*, in contrast, is non-material; biological science cannot illuminate its mode of operation. We face similar limitations when it comes to the universe. Only about 4% of what makes up the universe is normal matter; the rest is foreign to science. We therefore cannot expect to explain correctly the non-material aspect of the universe based exclusively on our knowledge of its material aspect. The scientific community is divided today because some modern scientists conflate the material and non-material aspects.

Making materialism the dead end of all investigations that involve material and non-material domains is unscientific and misleading. The consequences are that scientists (1) do not always follow the empirical evidence wherever it leads; (2) violate, if necessary, established scientific principles such as causality and laws such as biogenesis; and (3) reach conclusions that are contrary to our experiential knowledge and common sense.

According to information theory, genetic information is the difference between life and matter. To postulate that mutation (copying mistakes as raw material) and natural selection (a mindless process as the designing agent) are the sole reasons for the diversity and complexity of the material and non-material components of organisms is preposterous. For instance, Richard Dawkins says that genes wholly account for the complexity of human life. How can this be possible? Can enzymes create organs? It is unscientific to reach a conclusion based on preference and equivocal pieces of evidence.

The choice between an intelligent designer (God) and an unintelligent instrumentality (natural selection) is philosophical and not scientific. Choice of the former leads to theism, that of the latter to atheism. Since the choice is philosophical, scientists and lay persons are divided, both sides claiming their views to be consistent with the empirical evidence. In order to justify their choices, modern scientists seek to convince the public that their views are consistent with those of Albert Einstein and Sir Isaac Newton.

Einstein and Newton Were Not Evolutionists and Denigrated Atheism.

The Creation account described in Genesis was the dominant worldview before Darwin (1809-1882) introduced his controversial theory of evolution by natural selection. When there is a serious division in opinion, scientists invariably look for support from Newton (1642-1727) and Einstein (1879-1955). The objective of creationists as well as evolutionists is to claim that their views are consistent with those of these renowned giants of science. Small wonder, then, that Dawkins in *The God Delusion* devotes over five pages to distinguishing what he calls Einsteinian religion from supernatural religion, whereas with respect to Newton he merely notes:

> Newton did indeed claim to be religious. So did almost everybody until—significantly I think—the nineteenth century, when there was less social and judicial pressure than in earlier centuries to profess religion, and more scientific support for abandoning it.[7]

Newton's staunch belief in the Judaeo-Christian God made Newton the wrong model for evolutionist Dawkins, whose motive is to argue that God is a delusion. Dawkins' goal is also to convince the public that leading scientists are mostly atheists or agnostics, and he uses selective quotations to portray Einstein as such.

Dawkins states that Max Jammer's book *Einstein and Religion* is his main source of quotations from Einstein on religious matters.[8] This 250-page book contains numerous quotations, but Dawkins cites only those that are seemingly pro-atheist. For instance, Dawkins ignores statements by Einstein such as "I'm not an atheist, and I don't think I can call myself a pantheist."[9] In this sense Dawkins is unfair to his readers by misrepresenting Einstein's position. Interestingly, Einstein was aware of and responded to scholars such as Dawkins who used him to justify their philosophical preferences. The following excerpt from Jammer's book documents this fact:

> At a charity dinner in New York, Einstein explicitly dissociated himself from atheism when he spoke with the German anti-Nazi diplomat and author Hubertus zu Löwenstein: "In view of such harmony in the cosmos which I, with my limited human mind, am able to recognize, there are yet people who say there is no God. But what really makes me angry is that they quote me for support of such views.
>
> Recall his reaction to Büsching's book entitled *Es gibt keinen* Gott (There Is No God) in which he declared that a belief in a personal God seems "preferable to the lack of any transcendental outlook of life.[10]

To promote atheism and the evolutionist worldview, Dawkins omitted the above information, instead citing what others thought about Einstein. He writes:

> In greater numbers since his death, religious apologists understandably try to claim Einstein as one of their own. Some of his religious contemporaries saw him very differently. In 1940 Einstein wrote a famous paper

justifying his statement "I do not believe in a personal God." This and similar statements provoked a storm of letters from the religiously orthodox, many of them alluding to Einstein's Jewish origins.[11]

Dawkins goes on to quote statements made by a Roman Catholic bishop of Kansas City, an American lawyer working on behalf of an ecumenical coalition, a New York rabbi, and Reverend Dr. Fulton J. Sheen. He also presents letters from the president of an historical society in New Jersey and the founder of the Calvary Tabernacle Association in Oklahoma.

What is amusing is that Dawkins does not convey Einstein's response to all those who reacted to his paper. Max Jammer reports:

> Einstein described the reaction to his article quite caustically. "I was barked at by numerous dogs who are earning their food guarding ignorance and superstition for the benefit of those who profit from it. Then there are the fanatical atheists whose intolerance is of the same kind as the intolerance of the religious fanatics and comes from the same source. They are like slaves who are still feeling the weight of their chains which they have thrown off after hard struggle. They are creatures who—in their grudge against the traditional "opium for the people"—cannot bear the music of the spheres. The wonder of nature does not become smaller because one cannot measure it by the standards of human moral aims.[12]

Here again Einstein disassociates himself from both atheistic and religious fanatics. Clearly Dawkins contravenes Einstein's wish not to be counted among the atheists. Einstein believed that God exists, but he did not believe in a personal God who judges people's actions. "I do not believe in the God of theology who rewards good and punishes evil," he said. "My God created laws that take care of that. His universe is not ruled by wishful thinking, but by immutable laws."[13] What gives Dawkins the impression that Einstein's God is not supernatural? The fact is that Einstein believed in the existence of God, but like many Jews, Christians, and Moslems he did not believe in "a personal God who punishes the wicked or rewards the righteous and performs miracles by breaking the causal laws of nature."[14]

Although Einstein consistently maintained that he was not an atheist and indicated that he would prefer to believe in a personal God rather than to be an atheist, Dawkins obfuscates this fact:

> Much unfortunate confusion is caused by failure to distinguish what can be called Einsteinian religion from supernatural religion. Einstein sometimes invoked the name of God (and he is not the only atheistic scientist to do so), inviting misunderstanding by supernaturalists eager to misunderstand and claim so illustrious a thinker as their own.[15]

If, according to Dawkins, "Einstein was using 'God' in a purely metaphorical, poetic sense,"[16] why would he bother to declare that belief in a personal God seems preferable to atheism? Einstein's discontent with atheists is also evident in Walter Isaacson's article "Einstein and Faith" that appeared in *TIME* magazine: "[U]nlike Sigmund Freud or Bertrand Russell or George Bernard Shaw, Einstein never felt the urge to denigrate those who believe in God. Instead, he tended to denigrate atheists."[17] Jammer corroborates this view: "In spite of his denial of a personal God and his rejection of religious customs and rituals, he had a high respect for traditional religion."[18] The same author goes on to emphasize, "Einstein renounced atheism because he never considered his denial of a personal God as a denial of God."[19]

Former atheistic philosopher Anthony Flew stresses the point that "Dawkins cites Jammer on occasion, but leaves out numerous statements by Jammer and Einstein that are fatal to his case."[20] He writes:

> Another persistent theme in Dawkins' book, and in those of some of the other "new atheists," is the claim that no

scientist worth his or her salt believes in God. Dawkins, for instance, explains away Einstein's statements about God as metaphorical references to nature. Einstein himself, he says, is at best an atheist (like Dawkins) and at worst a pantheist. But this bit of Einsteinian exegesis is patently dishonest. Dawkins references only quotes that show Einstein's distaste for organized and revelational religion. He deliberately leaves out not just Einstein's comments about his belief in a "superior mind" and a "superior reasoning power" at work in the laws of nature, but also Einstein's specific denial that he is either a pantheist or an atheist.[21]

As far as a Darwinist like Dawkins is concerned, it is an unacceptable anomaly to be a great scientist while also being a theist.

For the scientific community at the time of Newton and Einstein, the aspiration was truth and understanding. Einstein remarked:

> [R]eligion determines the goal, science, in its broadest sense, shows the means for attaining this goal. However, "science can only be created by those who are thoroughly imbued with the aspiration toward truth and understanding. This source of feeling, however, springs from the sphere of religion . . . . I cannot conceive of a genuine scientist without that profound faith. The situation may be expressed by an image: science without religion is lame, religion without science is blind.
>
> A person who is religiously enlightened, appears to me to be one who has, to the best of his ability, liberated himself from the fetters of his selfish desires and is preoccupied with thoughts, feelings, and aspirations to which he clings because of their superpersonal value. What is important is the force of this superpersonal content . . . . regardless of whether any attempt is made to unite this content with a divine Being.[22]

Einstein, therefore, saw science and religion as two equally relevant aspect of knowledge. In the contemporary scientific community, the aspiration too often is a reductive and materialist answer to everything. Leading scientists today follow their philosophical presuppositions rather than the scientific evidence that points to God. Richard Lewontin, a leading evolutionist at Harvard University, thus writes:

> It is not that the methods and institutions of science somehow compel us to accept a material explanation of the phenomenal world, but, on the contrary, that we are forced by our *a priori* adherence to material causes to create an apparatus of investigation and a set of concepts that produce material explanations, no matter how counter-intuitive, no matter how mystifying to the uninitiated. Moreover, that materialism is absolute, for we cannot allow a Divine Foot in the door.[23]

Conclusions such as the evolution of life or information from inanimate matter, mind from mindless processes, and intelligence from random processes justify Einstein's previously cited thesis that "science without religion is lame." Einstein was only concerned with the truth and so he said, "I want to know how God created this world . . . . I want to know His thoughts."[24]

It should be evident from the preceding paragraphs that Einstein and Newton were not evolutionists. Both denigrated atheism and were consistent in their views that there is a God. They differed, however, in their concept of God. Einstein did not believe in a personal God, but Newton did and consequently wrote more volumes on religion than any other subject. Gale E. Christianson asserts:

> Besides containing invaluable keys for unlocking what Newton termed nature's "operations," Scripture to him served as both a moral guide and a means of determining what would happen in the future. The earliest clues to his private thoughts on God are contained in the confession

of sins he penned at Cambridge when he was nineteen: "Not turning nearer to Thee for my affections. Not living according to my belief. Not loving Thee for Thy self." From these simple sentences he would go on to write an estimated 1,400,000 words on religion, more than the alchemy, more than the mathematics, more even than the physics and astronomy that made him immortal.[25]

Of Sir Isaac Newton, Einstein remarked: "Nature to him was an open book, whose letters he could read without effort."[26] Here Einstein acknowledges Newton's unique gift for providing a foundation on which subsequent scientists could build. Creationists, in other words, established the foundation of modern science. It is thus hugely ironic that evolutionists today portray a belief in creationism as antithetical to the spirit of science.

Abiogenesis: Masking Science with Atheistic Assumptions.

Newton and Einstein did not regard science and religion as antithetical, and neither did the larger scientific community until Darwin presented his concept of natural selection as an alternative to supernatural design. With those two choices secularist scientists began to view science and religion as inherently antithetical. If Darwin's ideas on natural selection planted the seed of division within the scientific community, Louis Pasteur's empirical evidence on biogenesis split it into theists and atheists. Pasteur (1822-1895) affirmed the scientific law of cause and effect empirically and thus ruled out the myth that life can originate from dead matter (spontaneous biogenesis). Hubert Reeves et al., disappointed with the result of Pasteur's 1862 experiment, lament:

> Because of him [Pasteur], scientists concluded that life could not come directly from inert matter; therefore, it could come only from life itself. Which raised the essential question: how do you explain the initial manifestation of life? There were only three solutions: divine intervention, which removed the matter from the hands of science;

chance—in other words, some kind of accident—which took the matter into the realm of miracle, which is difficult to accept; or an extraterrestrial origin—germs of life that were brought here by meteorites—which didn't solve the question either.[27]

Since miracles are not part of science, the only logical conclusion is divine intervention, but such a conclusion is anathema to science.

In order to avoid the concept of God and promote atheism, therefore, modern scientists neglect the empirical evidence that suggests otherwise. This is nothing less than a deliberate abuse of science to promote a philosophical preference. The evolutionist worldview thus, is built on a myth. Nobel laureate George Wald wrote in 1954:

> We tell this story to beginning students of biology as though it represents a triumph of reason over mysticism. In fact it is very nearly the opposite. The reasonable view was to believe in spontaneous generation; the only alternative, to believe in a single, primary act of supernatural creation. There is no third position. For this reason many scientists a century ago chose to regard the belief in spontaneous generation as a "philosophical necessity." It is a symptom of the philosophical poverty of our time that this necessity is no longer appreciated. Most modern biologists, having reviewed with satisfaction the downfall of the spontaneous-generation hypothesis, yet unwilling to accept the alternative belief in special creation, are left with nothing. I think a scientist has no choice but to approach the origin of life through a hypothesis of spontaneous generation.[28]

Wald was wrong in suggesting that "a scientist has no choice but to approach the origin of life through a hypothesis of spontaneous generation." Thirty-eight years later Wald disclosed another option when he aligned his views with Einstein's:

In my life as scientist I have come upon two major problems which, though rooted in science, though they would occur in this form only to a scientist, project beyond science, and are I think ultimately insoluble as science. That is hardly to be wondered at, since one involves consciousness and the other, cosmology . . . . It has occurred to me lately—I must confess with some shock at first to my scientific sensibilities—that both questions might be brought into some degree of congruence. This is with the assumption that mind, rather than emerging as a late outgrowth in the evolution of life, has existed always as the matrix, the source and condition of physical reality—that the stuff of which physical reality is composed is mind stuff. It is mind that has composed a physical universe that breeds life, and so eventually evolves creatures that know and create: science-, art-, and technology-making animals.[29]

Other scientists agree. For instance, Nobel laureate George Snell identifies three areas where science fails to reveal ultimate truth: the nature of consciousness, the origins of design, and the matter of first causes.[30]

Wald abandoned his previous belief in abiogenesis because he upheld the fact that science must depend on empirical evidence and testable explanations confirming biogenesis. Meanwhile evolutionists doggedly stick to the unscientific premise of abiogenesis as a foundation for life forms and their diversity. Evolutionism thus divides the scientific community and compromises the discipline's integrity. Once scientists allow philosophical preference to prevail over established facts, they create a precedent for others to do the same. If you bend the rules for one topic, it is assumed that you can bend them for any topic.

The evolutionist worldview violates information theory and the second law of thermodynamics. According to this law, evolution should lead to downward regression as opposed to upward progression toward complexity, and this implies that "matter, on its own, does not organize itself to higher order."[31] The late A. E. Wilder-Smith stipulates:

> Neo-Darwinian thought requires basically the prebiotic autoorganization of raw matter (which the second law categorically excludes), the creation of information by random deviations (which information theory categorically forbids), the encoding of information by chance (without the help of exogenous code conventions), the storage of information by chance and its retrieval also by chance. The Darwinian hypothesis sets out to explain the origin and the replication of a biological organism (a super machine), immensely more complex than a modern automobile, by means of random deviations. If we were to accept such an hypothesis, we would have to be willing in principle to accept the origin and development of any other teleonomic machines solely and exclusively by means of the molecular deviations of iron molecules and by selection on the car market in the game of supply and demand, but without the aid of any teleonomic construction mechanisms, blueprints, or concepts . . . . Thus engineers, machines, and workshops would no longer be required to produce cars.[32]

Modern scientists have utilized unjustified assumptions and embraced nonsensical conclusions to explain away the scientific evidence that points to a superior mind. In doing so they simply fulfill this passage from the Scriptures:

> The truth about God is evident by instinct, made plain in our consciousness. For ever since the creation of the world, God's invisible nature and attributes are discernible. Some recognize this but respond negatively. They become godless in their thinking, coming up with vain imaginations, foolish reasoning and nonsense speculations. They choose not to honour God but instead gave credit to the things God created. They think of themselves

as being wise or smart, but in reality they are making fools of themselves. Though they exchange the truth of God for a lie and honoured the creature rather than the Creator, God is ever blessed forever, and never at risk of being defamed. (Rom. 1:19-25)

In fulfillment of the above prophecy, modern scientists have credited natural selection as a designing instrumentality equal to that of an intelligent Designer. Common sense dismisses such a conclusion as nonsensical. Little wonder, then, that Dawkins, expresses frustration with the public's reluctance to embrace Darwinism unreservedly: "It is almost as if the human brain were specifically designed to misunderstand Darwinism and to find it hard to believe."[33]

Evolution by natural selection cannot be proclaimed as a scientific fact concerning the origin of species when it cannot explain scientifically how life began in the first place. Empirical science clearly indicates that life cannot originate from dead matter (spontaneous generation). However, evolutionists, opting for a materialist paradigm, have ignored this hard evidence. This practice, in turn, has led to egregiously false conclusions.

DEBUNKING NATURAL SELECTION AS A CREATIVE AGENT

Douglas J. Futuyma describes the relevance of natural selection in biological science as follows:

> The theory of natural selection is the centerpiece of *The Origin of Species*. It is this theory that accounts for the adaptation of organisms, those innumerable features that so wonderfully equip them for survival and reproduction; it is this theory that accounts for the divergence of species from common ancestors and thus for the endless diversity of life . . . . Although it is merely a statement about rates of reproduction and mortality, the theory of natural selection is perhaps the most important idea in biology.

It is also one of the most important ideas in the history of human thought—"Darwin's dangerous idea," as the philosopher Daniel Dennett (1995) has called it—for it explains the apparent design of the living world without recourse to a supernatural, omnipotent designer.[34]

Similarly, Leslie Alan Horvitz provides this definition of natural selection: "The mechanism of evolution by which the environment acts on populations to enhance the adaptive ability and reproductive success of individuals possessing desirable genetic variants, increasing the chance that those beneficial traits will predominate in succeeding generations."[35] This mechanism, however, is inadequate to explain the origin and diversity of life forms.

While some scientists attach enormous importance to the doctrine of natural selection, others see it as mere tautology. Ex-evolutionist Gary Parker, for instance, contends that "natural selection is *not* some awesomely powerful scientific theory that enables scientists to predict future changes in population."[36] Parker further asserts:

Natural selection is a fact because it's a *tautology or truism,* a form of *circular reasoning. It is argued that the fittest are those that survive in greatest relative numbers, and those that survive in the greatest relative numbers are defined as the fittest.* That's definitely *true,* but it's really just an observation, not a profound theory, and begs the question of what makes some organisms fitter than others.[37]

There is nothing outstanding about natural selection, which is part and parcel of our daily life. We encounter some form of it, for example, in our digestive and respiratory systems. Natural selection only takes on an exaggerated meaning in evolutionary biology because of the false notion that it replaces God as the Designer. Scientists are divided on this issue. Evolutionist Richard Lewontin thus argues, "The manifest fit between organisms and their environment is a major outcome of evolution. Yet natural selection does not lead inevitably to adaptation; indeed, it is sometimes hard to define an adaptation."[38] As a result, all evolutionist assertions are speculative.

Agreeing with Parker, creationist Henry M. Morris describes natural selection as a "sort of sieve, through which pass the variants which fit the environment."[39] Natural selection only acts on variants; it does not create anything new. Morris concludes:

> Natural selection, acting upon the variational potential designed into the genetic code for each organism, is thus a powerful device for permitting horizontal variation, or radiation, to enable it adapt to the environment and thus to survive. It is useless, however, in generating a *vertical* variation, leading to the development of higher, more complex kinds of organism.[40]

Challenging the modern concept of macroevolution by natural selection, prominent evolutionist Stephen Jay Gould maintains that the belief that "all evolution is due to the accumulation of small genetic changes guided by natural selection" is essentially dead.[41] Here creationists concur with Gould. The creationism-evolutionism controversy is over macroevolution and not microevolution.

The unproductive and endless debates among evolutionists indicate that scientists are not confident in their understanding of how evolution occurs, and, this contradicts NASIM's official stance.[42] In support of Gould's view that many evolutionists question the role of natural selection, Steven M. Stanley in "A Theory of Evolution Above the Species Level" discredits the assertion that natural selection can account for the major features of evolution because the process "operates so slowly within established species."[43] This makes sense for a process that is random and blind. Stanley's argument is strengthened by the view of world-famous French zoologist and evolutionist Pierre-Paul Grassé:

> Mutations in time occur incoherently. They are not complementary to one another, nor are they cumulative

in successive generations toward a given direction. They modify what preexists, but they do so in disorder, no matter how.[44]

The evolutionist worldview derails if, according to Lewontin, natural selection does not necessarily lead to greater adaptation.[45] Therefore, although natural selection is a fact of science, it is an evolutionist *delusion* deployed to exclude God from the biological world.

Daniel R. Brooks and E. O. Wiley have undertaken a rigorous scholarly study of the evolution paradigm in light of entropy and information theory. In their study natural selection emerged simply as a minor component. Here are some highlights of their conclusions:

- Neo-Darwinism is a relatively complete theory of proximal causes in evolution . . . . Darwin referred to his theory as a theory of *natural selection*. In claiming that Darwinism is a complete theory of *evolution*, neo-Darwinists have unjustifiably extended proximal causes to the level of ultimate causes. This has produced an incomplete and relatively weak theory.

- The essential incompleteness of neo-Darwinism as a theory of evolution lies in its passive exclusion of axiomatic causal behaviour . . . . Our theory provides an axiom of ultimate causality, historically determined inherent directionality. The theory of natural selection nests within that more general theory of evolution as a theory of proximal causes affecting particular evolving lineages.[46]

Whether Brooks and Wiley are correct remains to be seen, but two important points stand out. First, their study is far from complete but more comprehensive than any other I have seen. Second, their study marginalizes the theory of natural selection as the creative force of evolution. It makes absolute sense because they incorporated both thermodynamics and information theory in their general theory.

Graeme Donald Snooks of Australian National University has recently challenged the theory of evolution by natural selection from the standpoint of a dynamic-strategy theory of life. This theory, unlike neo-Darwinism that deals with the way organic structures replicate themselves, focuses on the dynamics of life and human society. According to Snooks, his new realist model can explain the dynamics of both nature and the human society in a non-evolutionary framework. The key term in his theory is "biotransition" in place of evolution. In his book titled *The Collapse of Darwinism: or The Rise of a Realist Theory of Life*, Snooks recounts the failure of Darwinism and its variants to justify the need for a new approach:

> The neo-Darwinists have transformed natural selection from an economic concept involving the struggle for scarce resources into a sociological concept involving the struggle for "reproductive success." When taken to its logical conclusion by modern neo-Darwinists such as Edward Wilson and Richard Dawkins, the Darwinian focus shifts from organisms pursuing reproductive success to "selfish genes" attempting to maximize their presence in the gene pool by manipulating not only their host organism (or "survival machines") but also other organisms (or "extended phenotypes") as well. In doing so they have reduced Darwin's fatally flawed but serious concept into a subject of farce.[47]

Snooks continues his argument against the Darwinian theory by natural selection:

> The concept of natural selection is not a general dynamic theory, only an enabling device, and a faulty one at that. To begin with, its focus is all wrong. It is only possible to understand the dynamic process of genetic change at the individual, species, or dynasty level by

developing a general dynamic model of life in which it is but a component. Albeit an important component. This misfocus—encouraged by neo-Darwinism's exclusive concern with genetics—lies at the center of the failure of Darwinism. Although "evolution" is not the same as the dynamics of life, it is treated by the Darwinists as if it were. It is similar to developing a (faulty) model of technological change and then claiming that it can explain the dynamics of human society. There is no escaping the fact that the Darwinian model has no endogenous (or internal) driving force and no general dynamic mechanisms that can account for macrobiological change.[48]

Here Snooks describes the trivial role of natural selection as a designing instrumentality.

In his effort to counter creationist rhetoric, atheist Danny Vendramini isolates natural selection as a major problematical aspect of the NeoDarwinian theory. Vendramini contends that natural selection cannot create our predisposition to myths and, hence, proposes a secondary evolutionary process called Teemosis as the mechanism that "created and regulated many of the essential biosystems we are familiar with today—including emotions, memory, personality, attention, moods, perception, learning and motivation."[49] Atheist Jerry Fodor and Massimo Piattelli-Palmarini maintain that Darwin's theory of natural selection is fatally flawed.[50] "[O]rganisms can be modified and refined by natural selection, but that is NOT the way new species and new classes and new phyla originated."[51]

The emergence of such new theories and oppositions is evidence that the Darwinian concept of natural selection has been found wanting. A truly scientific depiction of the origins and diversity of organisms cannot posit natural selection as a creative force. We have considered Gould's discontent with the synthetic theory; Lewontin's assertion that natural selection does not lead to adaptation; Grassé's objections to the Darwinian and neo-Darwinian theory of evolution by natural selection; Brooks and Wiley's theory that insists natural selection cannot be a creative force; Snooks' realist theory of life in which the terms strategic selection and biotransition are substituted for that of natural selection and evolution respectively; Vendramini's 'teem theory' to account for natural selection deficiencies in biosystems; and Piattelli-Palmarini and Fodor's concern about the inadequacy of natural selection as an explanatory construct. The scholars I have discussed do not endorse God as the Supreme Designer, but they denigrate the role of natural selection as a creative instrumentality. However, many leading biologists such as Francisco Ayala and Richard Dawkins think otherwise. We shall next consider how these faithful disciples of Darwin propagate their belief in natural selection.

PROPAGATING THE FALLACY OF EVOLUTION BY NATURAL SELECTION

In comparing artificial to natural selection, Charles Darwin wrote:

> [M]an by selection can certainly produce great results, and can adapt organic beings to his own uses through the accumulation of slight but useful variations, given to him by the hand of Nature. But Natural Selection . . . is a power incessantly ready for action, and is as immeasurably superior to man's feeble efforts as the works of Nature are to those of Art.[52]

The above quotation is a fact if "Nature" is conscious and designed the objects for the selection process. It is a fallacy, however, if "Nature" is unconscious and works on objects pre-designed for the selection process. Darwin was apprehensive that the term "selection" implies conscious choice and that he speaks of natural selection as "an active power or Deity."[53] Darwin defines "Nature" as follows: "I mean by Nature only the aggregate action and product of many natural laws, and by laws the sequence of events as ascertained by us."[54] From Darwin's definition it is evident that Nature is unconscious. The belief that natural selection is "immeasurably superior" to artificial selection is, therefore, simply a delusion. This is not to deny the existence of natural selection as a process, but it *is* to refute its role as a *creative* power.

Neo-Darwinists continue to propagate misleading analogies for natural selection. Douglas J. Futuyma, for instance, maintains that "natural selection acts as an editor, not an author."[55] This begs the question, "Who is the author of the book called genetic mutation?" Futuyma, being an author himself, knows that the author of mutation is an external agent who consciously organizes materials to form a book. Chance, accidents, and unconscious or mindless processes do not qualify as authors. *An unconscious selector does not develop the material it selects.* Why then do evolutionists claim natural selection as a creative force?

In developing the theory of evolution by natural selection, Darwin relied heavily on an analogy between man's "methodical and unconscious means of selection" and natural selection. Snooks addresses this point:

> Darwin arrived at the concept of natural selection through what I have called the "farmyard" analogy. By assuming that organisms at all times and in all places are driven to maximize their number of offspring—what he called the "doctrine of Malthus"—Darwin was able to replace artificial selection in the farmyard with "natural selection," which he projected onto nature.[56]

Snooks continues:

> Darwin's analogy, while appearing appropriate at first sight, is totally misleading. Genetic change in the natural world is nothing like artificial breeding in the farmyard. When we subject Darwin's farmyard analogy to close scrutiny, it crumbles to dust.[57]

Artificial and natural selection are similar in the sense that both are selection processes and are limited in their operations. However, they are dissimilar in many respects. Norman Geisler and Peter Bocchino, in Table 6.1 below on artificial versus natural selection, demonstrate that the dissimilarities outweigh the similarities.[58]

Table 6:1 The Crucial Differences between Artificial and Natural Selection

	Artificial Selection	Natural Selection
Goal	Aim (end) in view	No aim (end) in view
Process	Intelligently guided process	Blind process
Choices	Intelligent choice of breeds	No intelligent choice of breeds
Protection	Breeds guarded from destructive forces	Breeds not guarded from destructive forces
Freaks	Preserves desired freaks	Eliminates most freaks
Interruptions	Continued interruption to reach desired goal	No continued interruption to reach any goal
Survival	Preferential survival	Nonpreferential survival

Darwin used the term "natural selection" in order to relate it to man's power of selection. From the above table it can be seen that natural selection mimics artificial selection but does so mindlessly and imperfectly. However, Darwin and his supporters, to promote their worldview, present natural selection as superior to artificial selection. Accordingly, Darwin stipulates:

> It may be said that natural selection is daily and hourly scrutinizing, throughout the world, every variation, even the slightest; rejecting that which is bad, preserving and adding up all that is good; silently and insensibly working, whenever and wherever opportunity offers, at the improvement of each organic being in relation to its organic and inorganic conditions of life. We see nothing of these slow changes in progress until the hand of time has marked the long lapses of ages, and then so imperfect is our view into long past geological ages that we only see that the forms of life are now different from what they formerly were.[59]

This is pure fantasy! In his rebuttal Snooks remarks:

> But these are only rhetorical flourishes. In reality natural selection is a passive filter that sorts out profitable from unprofitable variations and allows the former to accumulate slowly but continuously over vast periods of time. Natural selection, therefore, is not an active principle in life. That role is played by the assumed Malthusian propensity to populate and perish. By personifying an abstract filtering device rather than the organisms involved in the "war of nature," Darwin underlined the absence of any endogenous driving force in his system.[60]

This fact notwithstanding, evolutionists readily ascribe to natural selection the role of a designer. But natural selection, like artificial selection, can only modify the nature and not transform an organism into a different creature. Natural selection thus cannot be regarded as a designer but at best a modifier.

We next consider the role of natural selection as perceived from the perspective of an evolutionist who is a die-hard atheist

(Richard Dawkins) and that of an evolutionist who respects organized religion (Francisco J. Ayala).

Dawkins argues in *The Blind Watchmaker* that "Given infinite time, or infinite opportunities, anything is possible."[61] However, since Dawkins precludes the supernatural, he implies that some things are not possible even with infinite time. One of these impossible things is the infinite series of lucky events he invokes to explain his view of evolution:

> Cumulative selection is the key to all our modern explanations of life. It strings a series of acceptably lucky events [random mutations] together in a nonrandom sequence so that, at the end of the sequence, the finished product carries the illusion of being very very lucky indeed, far too improbable to have come about by chance alone, even given a timespan millions of times longer than the age of the universe so far . . . . So, cumulative selection can manufacture complexity while single-step selection cannot.[62]

Dawkin's "luck" is the equivalent of "miracle." Dawkins further argues:

> Cumulative selection, once it has begun, seems to me powerful enough to make the evolution of intelligence probable, if not inevitable . . . . The odds against assembling a well designed body that flies as well as a swift, or swims as well as a dolphin, or sees as well as a falcon, in a single blow of luck—single-step selection—are stupendously greater than the number of atoms in the universe, let alone the number of planets! No, it is certain that we are going to need a hefty measure of cumulative selection in our explanation of life.[63]

Just as Dawkins finds the story of creation impossible to believe, so his story of evolution by cumulative selection is far more impossible to believe. Lee M. Spetner offers this criticism:

"The events necessary for cumulative selection are much too improbable to build a theory on. The events needed for the origin of life are even more improbable."[64]

Dawkins substitutes genetic inheritance for the role of God, claiming that genes "created us, body and mind."[65] But how could genes *create* life? Here again, appealing to "lucky events," Dawkins concedes the emptiness of his atheistic speculations: "The account of the origin of life that I shall give is necessarily speculative; by definition, nobody was around to see what happened."[66] In this statement Dawkins acknowledges that his tale about human origin is not to be taken seriously as scientific fact and constitutes delusion. Other accomplished scientists, however, address the subject from a broader perspective.

Denis Noble, a physiologist and systems biologist at Oxford University, for instance, asserts that the genome or DNA does not define life. He equates Dawkins' "selfish-gene" view to a metaphorical polemic as opposed to "a straightforward empirical, scientific hypothesis."[67] Noble thus confirms the folk-tale nature of some of the key conclusions evolutionary biologists declare as scientific facts. He denounces the obsession with genes in the biological understanding of life and, instead, advocates a multi-level analysis comprising the interaction of processes from the molecular to organ, system, body, and environment. In order to justify his stance, Noble demonstrates below how different conclusions can be reached for the same scientific evidence on genes:

> Dawkins: [Genes] swarm in huge colonies, safe inside gigantic lumbering robots, sealed off from the outside world, communicating with it by tortuous indirect routes, manipulating it by remote control. They are in you and me; they created us, body and mind; and their preservation is the ultimate rationale for our existence.
>
> Noble: [Genes] are trapped in huge colonies, locked inside highly intelligent beings, moulded by the outside world, communicating with it by

> complex processes, through which, blindly,
> as if by magic, function emerges. They are in
> you and me; we are the system that allows
> their code to be read; and their preservation is
> totally dependent on the joy that we experience
> in reproducing ourselves. We are the ultimate
> rationale for their existence.[68]

In this exercise Noble compares Dawkins' metaphorical conclusion with a more realistic one. The scientific evidence is that "[Genes] are in you and me." Dawkins' and Noble's statements are the same in this regard since scientific facts are delusion-free. However, their conclusions differ. While Noble realistically points out that genes are a result of our existence, Dawkins incorrectly concludes that they are our creators. Dawkins' conclusion defies both common sense and experiential knowledge. Noble explains: "I see two major problems with it. The first is that, even if one does think that the genes code for a program that creates us, they certainly don't do that alone. The second is that I don't believe that such a program exists.[69]

Dawkins' conclusion that the genes in our bodies "created us, body and mind" is totally preconceived. He is simply using his academic position to lure people into atheism. Dawkins accordingly maintains that God is a delusion. However, he also understands that some things are beyond human comprehension and can be explained only by appealing to outside intelligence. So Dawkins recommends belief in extraterrestrial aliens rather than belief in God. Under the subheading "Little Green Men" he writes:

> Whether we ever get to know about them or not, there are
> very probably alien civilizations that are superhuman, to
> the point of being god-like in ways that exceed anything
> a theologian could possibly imagine . . . . The crucial
> difference between gods and god-like extraterrestrials lies
> not in their properties but in their provenance. Entities
> that are complex enough to be intelligent are products
> of an evolutionary process.[70]

Science has not been able to prove the existence of aliens; therefore, they are just as improbable as supernatural gods.

With the origin of life still under speculation, Dawkins tackles the question of species' diversity. He attributes this to the designing power of natural selection. Dawkins dismisses the teleological argument that points to God as the designer while affirming Darwin's conclusions:

> Thanks to Darwin, it is no longer true to say that nothing that we know looks designed unless it is designed. Evolution by natural selection produces an excellent simulacrum of design, mounting prodigious heights of complexity and elegance.[71]

Here again Dawkins waxes hyperbolic to credit the complexity and diversity of biological life to a natural process. In Dawkins' imaginary universe the blueprint is the raw material and an unconscious mechanism is the designer.

In *Evolution: A Theory in Crisis*, molecular biologist Michael Denton points to the lack of proof for Darwin's views (this applies to Dawkins' as well) on natural design:

> The Darwinian claim that all the adaptive design of nature has resulted from a random search, a mechanism unable to find the best solution in a game of checkers, is one of the most daring claims in the history of science. But it is also one of the least substantiated. No evolutionary biologist has ever produced any quantitative proof that designs of nature are in fact within the reach of chance.[72]

Dawkins' objective is to formulate a story of the origin and diversity of life in order to privilege Darwin's theory of evolution as engineered by natural selection. He therefore attempts to dismiss God's existence by using superficial probability arguments, but he is unable to provide a quantitative probability analysis to show why and how "Evolution by natural selection produces an excellent simulacrum of design."

We consider next the views of Francisco Ayala, a world-famous biologist who sees no conflict between religion and science. "Scientific knowledge," contends Ayala, "cannot contradict religious beliefs because science has nothing to say for or against revelation, religious realities, or religious value."[73] In the field of science, unlike Dawkins, Ayala points to natural selection as an imperfect mechanism as opposed to intelligent design:

> Evolution responds to the organisms' needs through natural selection not by optimal design but by "tinkering," by slowly modifying existing structures. Evolution achieves "design" as a consequence of natural selection while promoting adaptation. Evolution is "imperfect" design, rather than intelligent design.[74]

Ayala is unable to address, however, the origin of the raw materials (genes or DNA) upon which natural selection acts. And the truth is that no scientist knows. Without this knowledge, modern biology is grounded on myths. Denton thus concludes:

> The truth is that despite the prestige of evolutionary theory and the tremendous intellectual effort directed towards reducing living systems to the confines of Darwinian thought, nature refuses to be imprisoned. In the final analysis we still know very little about how new forms of life arise. The "mystery of mysteries"—the origin of new beings on earth—is still largely as enigmatic as when Darwin set sail on the *Beagle*.[75]

Therefore, if according to Dawkins belief in God is a delusion, belief in evolution by natural selection is a more subtle delusion.

Like any other selection process, nature imposes a limit on natural selection, but evolutionists are in denial for philosophical reasons. Novelties in biology are not produced by conjuring unsubstantiated probability. For novelties to emerge we require complex information that random natural processes cannot provide. A. E. Wilder-Smith asserts:

> There was one great aspect of reality about which Darwin—and indeed everyone of his epoch—knew nothing. I am referring to the modern science of information theory. For, if a primeval kind of amoeba is to develop up to a primate, that primeval cell will have to gather all sorts of new holistic information on how to make kidneys, livers, four chambered hearts, cerebra and cerebella, etc. For synthesis of such reduced entropy systems, as for example a primate brain, requires all kinds of solid actual holistic information which neither the matter of which the primeval amoeba consisted nor the intact amoeba cell contained. Similarly, inorganic matter will have to assemble huge numbers of bits of holistic information before it can synthesize an amoeba.[76]

Reorganization or a selection process can produce only a modified form of the original program, not a novel program with advanced features. Like artificial selection, irrespective of time, natural selection does not have the capacity to provide the additional information required to transform one kind of organism into a very different kind.

Ayala contends that "natural selection accounts not only for the preservation and improvement of the organization of living beings but also for their diversity."[77] The diversity that natural selection produces, just as artificial selection does, is confined to organisms' boundaries. Knowing that some people might not accept his scientific argument, Ayala maintains that "Imperfections and defects pervade the living world,"[78] which is why it is difficult to attribute the design of organisms to the biblical Creator. Ayala, therefore, adopts a position in which God is somewhere and ultimately nowhere.

Ayala's view that Darwin's theory of evolution is compatible with Christianity is admittedly flawed. Creationism and evolutionism are mutually exclusive paradigms, as are theism and atheism.

The Creation Sabbath Commandment unequivocally endorsed creation. Because evolutionism and creationism are exclusive as explanations of the origin and diversity of life on planet Earth, both cannot be both true.

WHY GOD IS PERSONAL

John Rennie argues that, in order to discredit an evolutionary worldview of life's origins, the Creator must appear before a live audience and claim credit.[79] Revelations through secondary sources, such as Genesis and other Creation stories, are inadequate. Victor J. Stenger, in *God: The Failed Hypothesis*, discusses some of these Creation narratives:

> An ancient Chinese myth tells us that everything started in chaos. The universe was like a black egg (a black hole?). A god named Pan Gu, wielding an axe, breaks the egg and the heavens begin to expand. The fleas and lice on Pan Gu's body evolve into humankind . . . . In the Bible and Qur'an, a presumably preexisting God creates the universe in six days. Following the story in Genesis, Earth is created on the first day. Four days later, God creates the sun, moon, stars.[80]

Stenger compares these creation myths with the Big Bang model, according to which the earth was not formed until nine billion years after that explosion.

> We see little resemblance in Genesis to the picture drawn by contemporary science . . . . All these facts can lead to only one conclusion: the biblical version of creation is dead wrong. The Chinese myth described above provides an account closer to the scientific one than the Bible's myth, picturing an expanding universe beginning in complete chaos and suggesting the evolution of life. However, it can hardly be considered an accurate description of the scientific data.[81]

The scientific model, however, can also be described as a myth since there is no proof that the universe was formed from a Big Bang. These myths, whether cultural, religious, or scientific, are outside the realm of science.

While the evolutionary accounts are myths, the Creation accounts can be established as fact through personal claims before a live audience. So far only the Judaeo-Christian God has fulfilled this requirement. The historical details of God's claims are reported in the Bible and in books on Jewish history. Further evidence is the seven-day weekly cycle that God imposed as a law, which remains in effect to this day. God personally communicates with people, professes love for them, and accepts their worship. In Chapter 9 I will give testimony that affirms a personal God, but here we must consider how Sir Isaac Newton's and Albert Einstein's views of God influence the creationism-evolutionism controversy.

Newton clearly believed in a personal God who receives worship, rewards virtue, and punishes evil conduct. Moreover, unlike Einstein, he endorsed biblical creationism. Einstein's disbelief in a personal God made it difficult for him to address the creationism-evolutionism debate. He therefore could only describe his concept of God and denigrate atheism. Both Newton and Einstein were creationists but differed in their view of God. From a pragmatic standpoint one of them is wrong.

Since God is an "illimitable superior spirit," as Einstein suggests and the Scriptures affirm (Gen. 1:2; Jer. 23:24), God can assume any form. Because the Scriptures contend that humanity is made in the image of God, it is logical, based on the principle of causality, to assert that *the first cause of personality must be personal*. Accordingly, "personality" is simply another attribute of God. In fact, Einstein's assertions that the "illimitable superior spirit who reveals himself . . . forms my idea of God"[82] and that "the divine reveals itself in the physical world"[83] corroborate this thesis. As an illimitable spirit God can choose to be anything except evil. In his argument against a personal God, Einstein limits the same God he describes as illimitable. This contradiction indicates that Newton's concept of an illimitable God with a personal attribute makes more sense. Above all, God must be personal if natural selection is to be refuted as a designing instrumentality.

Some modern scientists allow their emotions to influence their interpretation of the world around them. "The debate concerning creation/evolution," writes E. C. Ashby, "has not been entirely a scientific debate, but one involving positions argued by those who will or will not accept God as the creator of the universe and life, possibly because of the profound implications.[84] Such "profound implications" are the reason why evolutionism is generally preferred over creationism. Jeremy Rifkin has indicted the *hubris* implicit in evolutionism:

> We no longer feel ourselves to be guests in someone else's home and therefore obliged to make our behaviour conform with a set of pre-existing cosmic rules. It is our creation now. We make the rules. We establish the parameters of reality. We create the world, and because we do, we no longer feel beholden to outside forces. We no longer have to justify our behaviour, for we are now the architects of the universe. We are responsible to nothing outside ourselves, for we are the kingdom, the power, and the glory for ever and ever.[85]

In short, human beings often do not want to acknowledge God, whom they see as officious and punitive. Because Dawkins, for example, regards God as "thought-reading and sin-punishing,"[86] he endorses godlike creatures that in his mind are products of evolution. All of the above demonstrates that the primary objective of evolutionism is to eliminate God from our imagination.

Human beings developed science to understand the world around us. Therefore, science, as is true of human beings, is limited. Science cannot justify either theism or atheism. And the origin of species, like the origin of life, is not a scientific problem. Ariel Roth makes the following observation:

> Science made its greatest error when it rejected God and everything else except mechanistic explanations. By

failing to recognize its limitations, science has attempted to answer almost everything within a purely naturalistic philosophy. Evolution then became the most plausible model of origins. Science would not now be facing apparently insurmountable challenges to evolution if it had not adopted such a strong, exclusive, naturalistic stance . . . . The problem is not just evolution. In a sense, evolution is only an important symptom of a more deep-seated issue. The real difficulty is more whether naturalistic science is going to persist in trying to provide answers to all questions within its own closed system of explanations. How did science get into this intellectual straightjacket?[87]

The goal of science is to provide veridical explanations and answers within its own domain and without unprovable presuppositions.

NOTES

1. "Social Responsibility and the Scientist,", *Perspectives in Biology and Medicine in Modern Western Society* 14 (1971), p. 367.

2. *Origins: Linking Science and Scripture* (Hagerstown, MD: Herald Publishing Association, 1998), p. 182.

3. *Evolution of Living Organisms: Evidence for a New Theory of Transformation* (New York: Academic Press, 1977), p. 8.

4. Quoted in Max Jammer, *Einstein and Religion: Physics and Theology* (Princeton: Princeton University Press, 1999), p. 123.

5. "Changing Man," *Science* 155 (1967), pp. 409-10.

6. *Science and Creationism: A View from the National Academy of Sciences*, 2nd ed. (Washington, D.C.: National Academy Press, 1999), p. 10.

7. *The God Delusion* (Boston: Houghton Mifflin, 2006), p. 98.

8. Ibid, p. 16.

9. *Einstein and Religion*, p. 48.

10. Ibid, pp. 96-97, 149.

11. *The God Delusion*, pp.15-16.

12. *Einstein and Religion*, p. 97.

13. Ibid, pp. 122-23.

14. Ibid, p. 149.

15. *The God Delusion*, p. 13.

16. Ibid, p. 18.

17. "Einstein and Faith," *Time,* 16 April 2007, pp. 35-36.

18. *Einstein and Religion*, p. 149.

19. Ibid, p. 150.

20. *There Is a God: How the World's Most Notorious Atheist Changed His Mind* (New York: HarperCollins, 2007), p. 100.

21. Ibid, pp. xxii-xxiii.

22. *Einstein and Religion*, pp. 93-94.

23. "Billions and Billions of Demons," *The New York Review*, 9 January 1997, p. 31.

24. *Einstein and Religion*, p. 123.

25. *Isaac Newton* (New York: Oxford University Press, 2005), p. 60.

26. Quoted in Gale E. Christianson, *Isaac Newton and the Scientific Revolution* (New York: Oxford University Press, 1996), p. 144.

27. Hubert Reeves, Joel De Rosnay, and Yves Coppens, *Origins: Cosmos, Earth, and Mankind* (New York: Arcade Publishing, 1998), p. 78.

28. "The Origin of Life," *Scientific American* 191.2 (1954), p. 46.

29. "Life and Mind in the Universe," *Cosmos, Bios, Theos*, ed. Henry Margenau and Roy Abraham Varghese (La Salle, IL: Open Court, 1992), pp. 218-19.

30. "I Do Not See How Science Can Shed Light on the Origins of Design," *Cosmos, Bios, Theos*, p. 209.

31. A. E. Wilder-Smith, *The Natural Sciences Know Nothing of Evolution* (Costa Mesa, CA: T.W.F.T. Publishers, 1981), p. vi.

32. Ibid, pp. 72-73.

33. *The Blind Watchmaker* (New York: Penguin, 2006), p. xix.

34. *Evolution* (Sunderland: Sinauer Associates,2005), p. 247.

35. *The Complete Idiot's Guide to Evolution* (Indianapolis: Alpha Books, 2002), p. 303.

36. *Creation: Facts of Life* (Green Forest, AR: Master Books, 2006), p. 87.

37. Ibid.

38. "Adaptation," *Scientific American* 239 (1978), p. 213.

39. *Scientific Creationism* (Green Forest, AR: Master Books, 1998), p. 52.

40. Ibid, p. 53.

41. "Darwinism and the Expansion of Evolutionary Theory," *Science* 216 (1982), p. 382.

42. See, for example, *Science and Creationism*, p. xiii.

43. "A Theory of Evolution Above the Species Level," *Proceedings: National Academy of Sciences* 72.2 (1975), p. 646.

44. *Evolution of Living Organisms*, p. 97.

45. "Adaptation," p. 222.

46. *Evolution as Entropy: Toward a Unified Theory of Biology* (Chicago: University of Chicago Press, 1988), pp. 367-68.

47. *The Collapse of Darwinism: or The Rise of a Realist Theory of Life* (New York: Lexington Books, 2003), pp. 10-11.

48. Ibid, pp. 11-12.

49. Vendramini D., "The Second Evolution: The unified Teem Theory of Evolution, Perception, Emotions, Behaviour and Inheritance", http://thesecondevolution.com/Introduction.pdf. Retrieved on 3 April 2011.

50. *What Darwin Got Wrong*.(New York: Farrar, Straus and Giroux, 2010), p. xiv.

51. Mazur, S., "Piattelli-Palmarini: Ostracism W/out Nat Selection" http://www.scoop.co.nz/stories/print.html?path=HL0805/S00106.htm Posted Friday, 9 May 2008. Retrieved on 8/13/2008.

52. *The Origin of Species* (New York: Penguin, 1985), p. 115.

53. Ibid, p. 89.

54. Ibid

55. *Science on Trial: The Case for Evolution* (New York: Pantheon, 1995), p. 223.

56. *The Collapse of Darwinism*, p. 9.

57. Ibid, p. 33.

58. *Unshakable Foundations: Contemporary Answers to Crucial Questions about the Christian Faith* (Minneapolis: Bethany House, 2000), p. 148.

59. *The Origin of Species*, p. 133.

60. *The Collapse of Darwinism*, pp. 27-28.

61. *The Blind Watchmaker*, p.139.

62. Ibid, pp. 139-41.

63. Ibid, p. 146.

64. *Not by Chance: Shattering the Modern Theory of Evolution* (New York: Judaica Press, 1997, p. 166.

65. *The Selfish Gene* (Oxford: Oxford University Press, 2006), pp.19-20.

66. Ibid, p. 14.

67. *The Music of Life*: *Biology Beyond the Genome* (Oxford: Oxford University Press, 2006), p. 11.

68. Ibid, p. 12.

69. Ibid, p. 20.

70. *The God Delusion*, p. 258.

71. Ibid, p. 79.

72. *Evolution: A Theory in Crisis* (Bethesda, MD: Adler and Adler, 1986), p. 324.

73. *Darwin and Intelligent Design* (Minneapolis: Fortress Press, 2006), p. 103.
74. Ibid, pp. 86-87.
75. *Evolution: A Theory in Crisis*, pp. 358-59.
76. *The Scientific Alternative to Neo-Darwinian Evolutionary Theory* (Costa Mesa, CA: Word for Today, 1987), p. xii.
77. *Darwin and Intelligent Design*, p. 58.
78. Ibid, p. 85.
79. "15 Answers to Creationist Nonsense," *Scientific American* 287.1 (2002), p. 80.
80. *God: The Failed Hypothesis. How Science Shows That God Does Not Exist* (New York: Prometheus Books, 2007), p. 174.
81. Ibid, pp. 174-75.
82. "Obituary: Einstein Noted as an Iconoclast in Research, Politics, and Religion," *New York Times*, 19 April 1955, p. 25.
83. Quoted in Jammer, *Einstein and Religion*, p. 151.
84. *Understanding the Creation/Evolution Controversy* (Ozark, AL: ACW Press, 2005), p. 108.
85. *Algeny: A New Word—A New World* (New York: Viking, 1983), p. 244.
86. *The God Delusion*, p. 19.
87. *Origins: Linking Science and Scripture*, p. 334.

CHAPTER 7

THE FOSSIL RECORD DELUSION

Creation and evolution, between them, exhaust the possible explanations for the origin of living things. Organisms either appeared on the earth fully developed or they did not. If they did not, they must have developed from preexisting species by some process of modification. If they did appear in a fully developed state, they must indeed have been created by some omnipotent intelligence, for no natural process could possibly form inanimate molecules into an elephant or a redwood tree in one step.[1]

—Douglas J. Futuyma

In any local area, a species does not arise gradually by the steady transformation of its ancestors: it appears all at once and "fully formed."[2]

—Stephen Jay Gould

In any case, no real evolutionist, whether gradualist or punctuationist, uses the fossil record as evidence in favour of the theory of evolution as opposed to special creation.[3]

—Mark Ridley

THE FOSSIL RECORD:
THE NATURAL DOCUMENT OF CREATION

God claimed credit for having created the universe and documented this claim on stone tablets. The fossil record corroborates God's claim. French zoologist Pierre-P. Grassé insists:

> Naturalists must remember that the process of evolution is revealed only through fossil forms. A knowledge of paleontology is, therefore, a prerequisite; only paleontology can provide them with the evidence of evolution and reveal its course or mechanisms. Neither the examination of present beings, nor imagination, nor theories can serve as a substitute for paleontological documents. If they ignore them, biologists, the philosophers of nature, indulge in numerous commentaries and can only come up with hypotheses. This is why we constantly have recourse to paleontology, the only true science of evolution. From it we learn how to interpret present occurrences cautiously; it reveals that certain hypotheses considered certainties by their authors are in fact questionable or even illegitimate.[4]

Arguments based on morphological similarity cannot establish whether species are products or creation or evolution.

Carl O. Dunbar in *Historical Geology* describes the case for evolution as follows: "While the comparative study of living animals and plants may give very convincing circumstantial evidence, fossils provide the only historical, documentary evidence that life has evolved from simpler to more complex forms."[5] Although evolution can be verified only through fossil forms, creation is revealed both by fossil forms as well as by the Creator's claim. Since the idea of a divine Creator is anathema to science, it comes down to the question of which worldview the fossil evidence supports.

From a single fossil fragment, such as a tooth or skull, a paleontologist cannot tell whether an organism is the product of creation or evolution. Controversies emerge when scientists try to build an evolutionary history around a single fossil of an organism such as Archaeopteryx or Tiktaalik. Under what conditions, then, can scientists say whether a fossil evolved or not?

Since the geological cross-section of the earth is not consistent, the only reliable proof of either evolution or creation is the examination of a collection of fossil forms from different organisms within the same geological location. Fossils that are fully formed with no transitional gradations will characterize a collection that is consistent with the creation account in the book of Genesis. If, on the other hand, the designing instrumentality is natural selection, the collection will consist of basic fossil forms plus numerous transitional gradations. Therefore, we need locations where there is an abundance of fossil forms in order to establish the case for either creation or evolution.

Two locations have an abundance of early fossils: one is the Burgess Shale in the Canadian Rockies, and the other is the Chengjiang site in China. This great quantity of varied life forms is found in the geological column usually described as the "Cambrian Explosion." While the Cambrian layer discloses virtually every phylum known to man, the layers below reveal almost nothing about fossilized specimens; in the layers above the number of species fossilized gradually decrease with each successive layer, and at the most recent layers approximately 98 percent of every thing that has ever lived is extinct.[6] Thus, "contrary to the tree of life depicted in the school books, the fossil record depicts exactly the opposite story[.] The tree of life is an inverted cone, and not a tree at all."[7] Small wonder, since it is a fictitious construct. To show their disdain, "Evolutionists have come up with just about every explanation for the Cambrian explosion *except* the biblical model."[8]

The Cambrian Explosion fits the creationist worldview. The major types of organisms appear fully formed with no intermediate transitions. In fact, Darwin considered this sudden appearance of species belonging to several of the main divisions of the animal kingdom with no antecedents to be the greatest single objection to his theory of evolution by natural selection.[9] Half a century after *The Origin of Species* was published, Austin H. Clark of the U.S. National Museum made the following remarks:

- No matter how far back we go in the fossil record of previous animal life upon the earth we find no trace of any forms which are intermediate between the various major groups or phyla.
- This can only mean one thing. There can be only one interpretation of this entire lack of any intermediates between the major groups of animals—as for instance between the backbone animals or vertebrates, the echinoderms, the mollusks, and the arthropods.
- If we are willing to accept the facts we must believe that there never were such intermediates, or in other words that these major groups have from the very first borne the same relation to each other that they bear today.[10]

Die-hard evolutionists are reluctant to accept any evidence that is antithetical to their worldview.

Based on extant scientific evidence, we therefore conclude that organisms did not develop from preexisting species by some process of modification. Since the species are all found in a fully developed state, they are the products of creation by an omnipotent intelligence, in accordance with this chapter's first epigraph by evolutionist Douglas J. Futuyma.

The initial expression of frustration with the fossil data came from Darwin, the father of evolution by natural selection. He remarks:

> The main cause, however, of innumerable intermediate links not now occurring everywhere throughout nature depends on the very process of natural selection, through which new varieties continually take the places of and exterminate their parent-forms. But just in proportion as this process of extermination has acted on an enormous scale, so must the number of intermediate varieties, which have formerly existed on the earth, be truly enormous. Why then is not every geological formation and every stratum full of such intermediate links? Geology assuredly does

not reveal any such finely graduated organic chain; and this, perhaps, is the most obvious and gravest objection which can be urged against my theory.[11]

Darwin blamed the lack of intermediate varieties on the imperfection of the geological record and hoped that with time some transitional fossils would be uncovered. Well over a century after Darwin's time, paleontologist Stephen Jay Gould notes the lack of transitional evidence and recalls Darwin's reluctance to accept the truth:

> The extreme rarity of transitional forms in the fossil record persists as the trade secret of paleontology. The evolutionary trees that adorn our textbooks have data only at the tips and nodes of their branches: the rest is inference, however reasonable, not the evidence of fossils. Yet Darwin was so wedded to gradualism that he wagered his entire theory on a denial of this literal record. The geological record is extremely imperfect, and this fact will to a large extent explain why we do not find interminable varieties, connecting together all the extinct and existing forms of life by the finest graduated steps. He who rejects these views on the nature of the geological record, will rightly reject my whole theory.[12]

The Darwinian theory of evolution by natural selection thus flunks the only reliable and authentic test that would validate its claims.

Unwilling to abandon Darwin's theory of evolution, some scientists have looked for possible solutions to sustain it. Accordingly, Gould and Niles Eldredge of the American Museum of Natural history developed the theory of punctuated equilibrium.[13] Things still did not improve much, as we learn from Michael Denton's analysis of the fossil record:

- Only a small fraction of the hundred thousand or so fossil species known today were known to Darwin. But virtually all the new fossil species discovered since Darwin's time have either been closely related to known forms or, like the Poganophoras, strange unique types of unknown affinity.

- The fossils have not only failed to yield the host of transitional forms demanded by evolution theory, but because nearly all extinct species and groups revealed by paleontology are quite distinct and isolated as they burst into the record, then the number of hypothetical connecting links to join its diverse branches is necessarily greatly increased.

- Because the soft biology of extinct groups can never be known with certainty, then obviously the status of even the most convincing intermediates is bound to be insecure . . . . It is possible to allude to a number of species and groups such as *Archaeopteryx*, or the rhipidistian fish, which appear to be to some extent intermediate. But even if such were intermediate to some degree, there is no evidence that they are any more intermediate than groups such as the living lungfish or monotremes which, as we have seen, are not only tremendously isolated from their nearest cousins, but which have individual organ systems that are not strictly transitional at all. As evidence for the existence of natural links between the great divisions of nature, they are only convincing to someone already convinced of the reality of organic evolution.[14]

Here Denton makes an honest scientific evaluation of the situation. He distances himself from other evolutionary biologists who concede that there are serious problems with the Darwinian model but believe that they can be explained away by making only minor adjustments to the framework. These Darwinian devotees point to a handful of fossils as evidence in favour of the theory and blame the rest on the imperfection of the fossil record. For instance, Ernst Mayr in *What Evolution Is,* written almost two and a half decades after the theory of punctuated equilibrium was introduced, addresses the fossil problem as follows:

> Given the fact of evolution, one would expect the fossils to document a gradual steady change from ancestral forms to the descendants. But this is not what the paleontologist finds. Instead, he or she finds gaps in just about every phyletic series. New types often appear quite suddenly, and their immediate ancestors are absent in the earlier geological strata . . . . Indeed the fossil record is one of discontinuities, seemingly documenting jumps (saltations) from one type of organism to a different type. This raises a puzzling question: Why does the fossil record fail to reflect the gradual change one would expect from evolution?[15]

In his explanation Mayr reinforces Darwin's argument about the "unimaginable incompleteness" of the fossil record and cites cases of fossils that appear to be intermediate between reptiles and mammals.[16] However, the fossils to which Mayr alludes, such as *Archaeopteryx* discovered in Germany and *Australopithecine* discovered in South Africa, came from different geological sites. Analysis of isolated fossils from different sites cannot tell us anything concrete about their origins or life history. We need an accumulation of fossils from a given location, such as the Chengjiang site in China, to reach a correct conclusion. We now will consider three fossils that have captured the world's attention in order to understand why claims made by evolutionists are disputable.

THE MISSING LINKS:
TOUMAI, TIKTAALIK, AND PLATYPUS

The term "missing link," which according to Ernst Mayr is "a fossil bridging the large gap between an ancestral and a derived group of organisms,"[17] generates another controversy. The current definition presumes that Darwinian evolution has occurred. For instance, the second edition of *The Oxford English Reference Dictionary* (1996) defines "missing link" as follows:

> [A] hypothetical intermediate type, especially between humans and apes. The missing link was a Victorian concept, arising from a simplistic picture of human evolution, and represented either a common evolutionary ancestor for both humans and apes, or, in popular thought, some kind of ape-man through which humans had evolved from the other higher primates.[18]

Although standard dictionaries define "missing links" as "hypothetical," evolutionists regard them as real. It is therefore important to provide a rigorous definition in order to address properly the creationism-evolutionism controversy. We should be able to know when an organism is a real missing link or an imaginary construct.

An organism is missing if it has been physically observed and/ or documented at some point in time but is no longer evident. Accordingly, in science as opposed to pseudoscience, a fossil can only be a real "missing link" when an intermediate variation currently exists or is documented as having once existed. For instance, platypus can be so classified. If, on the other hand, an organism's existence at some point in history is simply assumed based on limited data, such fossils can be described only as hypothetical or imaginary missing links. Toumai and Tiktaalik belong to this category. A fossil that is in abundance but no longer found in the living world, such as dinosaurs, is described as extinct.

Disputes continue among evolutionists as to whether a given fossil is a missing link or not. For instance, concerning Toumai, the fossilized skull discovered in Chad that was headline news in July 2002, *TIME* magazine reported that the interpretation of this evidence depended largely on which paleontologist you asked.[19] In October 2002 two articles with divergent views were published in *Nature*. One claimed that Toumai was an ape, the other that it was a modern human ancestor.[20] Here we are dealing not with real but with imaginary missing links.

Four years thereafter evolutionists captured the world's attention with the discovery of the fossil Tiktaalik, which NASIM contends is "a notable transitional form between fish and the early tetrapods that lived on land."[21] This is an assumption, however, not a scientific fact. As defined above, Tiktaalik is simply an imaginary missing link since in the living world no organism closely resembles it.

On 8 May 2008, the same year in which NASIM published its booklet *Science, Evolution, and Creationism,* a global two-year study released an analysis of the platypus genome.[22] The genome reveals, in addition to unique platypus characteristics, modules that are partly avian, partly reptilian, and partly mammalian. Below are some highlights of the genome's features:

- It is about two-thirds the size of the human genome; composed of 18,500 genes like other vertebrates; and furnished with 52 chromosones, including 10 sex chromosomes. The platypus is a monotreme characterized by a common opening for the excretion of faeces, urine, and for the passage of eggs or sperm. It uses a complex electrosensory system to forage for food underwater.
- It possesses avian qualities implied by its Latin name, *Ornithorhynchynchus anatinus.* These include some bird-like microRNAs, a sex chromosome very similar to the bird Z sex chromosome; a duck-like flat bill and webbed feet.
- It has reptilian attributes; the female lays eggs; and the male's hind feet are equipped with a spur that discharges snake-like venom.
- It has mammalian characteristics such as genes that produce milk, contains mammal—like microRNAs, and is covered in fur. It has no nipples, but the young feeds through tiny openings located in the skin of the mother's belly.

The above are scientific facts. However, whether the multifaceted features of the platypus are evidence for evolution or creation is largely an opinion. In accordance with Feynman's rule for scientific integrity, both possibilities should be addressed in order for the public to understand the significance of this research.

Evolutionists would describe the platypus as a missing link, like Tiktaalik, if it were found only in the fossil record. Contrary to evolutionist propaganda about the usefulness of its genome in understanding mammalian evolution, however, the platypus is evidence that opposes evolutionism. Proponents have no option but to shy away from the embarrassing presence of platypus as a living organism. The empirical evidence suggests that the platypus defies the classic Darwinian view of biological evolution by small cumulative changes and the influence of natural selection. Massimo Piattelli Palmarini at the University of Arizona expressed this defeat as "Platypus one, Darwin zero."[23]

It is important that we understand why evolutionists are gravely worried about the evolutionary status of the platypus. If we remain faithful to the raw genome data, which suggest that the platypus is an amalgam of avian, reptilian, and mammalian features, then the evolutionary tree of life in its current form is wrong. The estimated time of origin of reptiles is 370 million years and of birds and mammals is 225 million years ago.[24] If the platypus is connected to reptiles, birds and mammals, then it should be found in strata 370-225 million years old, but the oldest platypus fossil is only about 120 million years old.[25] Evolutionists must begin to admit that they are wrong, as Robert Carter suggests:

> Even evolutionists need to concede that we are not in fact sampling an evolutionary tree but the terminal branches. Indeed, there is little to no evidence that any of the evolutionary nodes (missing links) ever existed. Try as we might, we cannot get into the fossil record to study the genetics of the nodes.[26]

Accordingly, scientists who adhere to the empirical evidence are now publicly discussing the uprooting of Darwin's tree of life.[27] This, of course, has been long overdue.

There is, however, one important argument that has arisen from the platypus genome project: the view that the platypus is an evolutionary missing link.[28] Darwin referred to platypus (*Ornithorhynchus*) as a living fossil.[29] Based upon the scientific evidence, about 121 million years ago the early platypus "had already developed features thought to be unique to modern platypuses, including an electro-sensitive 'bill' for finding aquatic prey."[30] The platypuses, therefore, except for variations in size and the absence of teeth in modern adults, have not changed significantly. Scientists have found no evidence of a "diversity explosion" at any time in monotreme history.[31] Differences in size and the absence of teeth are minor variations that cannot be regarded as valid proof of evolutionary history leading to complexity. These features are merely reflective of artificial breeding in which breeders produce, say, for instance, grape fruits that are smaller or bigger in sizes than normal ones and those with or without seeds.

Michael Denton argues that, if the various anatomical and physiological systems of monotremes were strictly transitional between reptiles and mammals, the case for their being genuine transitional types would be much clearer. "Instead of finding character traits which are obviously transitional," however, "we find them to be either basically reptilian or basically mammalian, so that although the monotremes are a puzzle in terms of typology they afford little evidence for believing that any of the basic character traits of the mammals were achieved gradually in the way evolution envisages."[32] The history of platypus thus does not lend support to an evolutionist worldview.

From the preceding discussion follow these logical conclusions:

1. Platypus is found both in the fossil record and among living organisms. It comes fully formed with no evidence of transitional or intermediate stages. Moreover, since genome analysis of the platypus reveals unique signatures confirming its external features as part bird, mammal, and reptile, platypus is one of the basic or ancestral living organisms. For almost two centuries evolutionists have

been searching for missing links to justify evolutionism. Platypus is evidence of the false notion of an "evolutionary missing link."

2. Platypus has unique features such as an electro-sensitive bill and venom. These characteristics further support the conclusion that the platypus is a uniquely designed organism as opposed to an evolved organism. As such, then, the empirical data does not fit the theoretical evolutionary tree described in modern science textbooks.

3. A comparison of the modern platypus with ancient fossils reveals only intraevolutionary (microevolutionary) changes as opposed to extraevolutionary (macroevolutionary) changes. This is consistent with the extant evidence of other basic kinds of organisms that appear abruptly in the fossil record without alleged transitions. *The only logical conclusion is that there are no "missing links" in the history of species, whether in the fossil record or the living world.*

Evolutionism and creationism, of course, reach different conclusions based on the same scientific evidence. The scientific facts in this case are (1) the fossil record of platypus and (2) genome data on the platypus as alive today. When these data are processed without reference to either biblical creationism or secular evolutionism, we notice that the platypus is a unique organism whose biological life history has not progressed from simple to complex over the ages. To reach a conclusion about whether the platypus evolved or was created, we must seek additional information.

Futuyma argues that if organisms "did appear in a fully developed state, they must indeed have been created by some omnipotent intelligence, for no natural process could possibly form inanimate molecules into an elephant or a redwood tree in one step."[33] Paleontologist Gould maintains that "a species does not arise gradually by the steady transformation of its ancestors: it appears all at once and fully formed."[34] In addition, evolutionary biologist Mayr confirms that paleontologists find "gaps in just about every phyletic series; new types often appear quite suddenly, and their immediate ancestors are absent in the earlier geological strata."[35] No matter how compelling pieces of circumstantial evidence are, they will never suffice as proof that bacteria-to-human evolution has occurred. Therefore, we may conclude that organisms did not evolve by chance. The Judaeo-Christian God affirms this conclusion by His claim, "It is I who made the earth and created mankind upon it. My own hands stretched out the heavens: I marshalled their starry hosts" (Isa. 45:12 NIV).

Regarding the fossil record, evolutionist E. J. H. Corner, a Cambridge University botanist, reaches the same conclusion as zoologist Mark Ridley of Oxford University. Corner writes: "Much evidence can be adduced in favour of the theory of evolution—from biology, bio-geography and palaeontology, but I still think that, to the unprejudiced, the fossil record of plants is in favour of special creation."[36] While some anti-creationists acknowledge scientific facts contrary to the evolutionist worldview, others insist that "absence of evidence is not necessarily evidence of absence."[37] NASIM here is in the forefront. Its members are mostly evolutionists who, in an attempt to privilege evolutionism, present only scientific evidence that supports its paradigm.

It is rather pathetic that hardened evolutionists cannot accept the shortcomings of their theory. For instance, according to Richard Dawkins, "Evolution by natural selection produces an excellent simulacrum of design, mounting prodigious heights of complexity and elegance."[38] This is a very strong claim. In order for it to be true, we expect transitional organisms produced by natural selection to constitute over fifty percent of the fossil record. What we find, however, is that we cannot rule in favour of evolution from isolated fossils excavated from different locations. It is also impossible to reconstruct the exact morphologies of organisms that lived in the past by examining a few pieces of their skeletal remains. All things considered, the fossil record supports a creationist worldview.

NOTES

1. *Science on Trial: The Case for Evolution* (New York: Pantheon, 1983), p. 197.
2. "This View of Life: Evolution's Erratic Pace," *Natural History* 86.5 (1977), p. 14.
3. "Who Doubts Evolution?" *NewScientist* 90 (1981), p. 831.
4. *Evolution of Living Organisms* (New York: Academic Press, 1977), p. 4.
5. *Historical Geology*, 2nd ed. (New York: John Wiley, 1960), p. 47.
6. See "Cambrian Explosion Disproves Evolution." http://www. learnthebible.org/ creation_science_cambrian_explosion_ disproves_evolution.htm.
7. Ibid.
8. Gary Parker, *Creation: Facts of Life* (Green Forest, AR: Master Books, 2006), p. 158.
9. See *The Origin of Species* (New York: Signet, 2003), p 315.
10. *The New Evolution Zoogenesis* (Baltimore: Williams and Wilkins, 1930), p. 189.
11. *The Origin of Species* (New York: Penguin, 1985), pp. 291-92.
12. "This View of Life: Evolution's Erratic Pace," p. 14.
13. "Punctuated Equilibria: The Tempo and Mode of Evolution Reconsidered," *Paleobiology* 3.2 (1977), pp. 115-51.
14. *Evolution: A Theory in Crisis* (Bethesda, MD: Adler and Adler, 1986), pp. 160-61, 165-66, 194-95.
15. *What Evolution Is* (New York: Basic Books, 2001), p. 14.
16. Ibid, pp. 14-15.
17. Ibid, p. 288.
18. *The Oxford English Reference Dictionary*, 2nd ed. (Oxford: Oxford University Press, 1996), p. 924.
19. Michael Lemonick, Andrea Dorfman, and Delphine Schrank, "Father of Us All?" *TIME*, 22 July 2002, p. 38.
20. Milford H. Wolpoff, Brigitte Senut, Martin Pickford, John Hawks, "Sahelanthropus or 'Sahelpithecus'?" and Michel Brunet, et al., "Reply," *Nature*, 10 October 2002, pp. 581-82.

21. *Science, Evolution, and Creationism* (Washington, D.C.: National Academies Press, 2008, p. 23.

22. W. C. Warren, L. W. Hillier, J. A. M. Graves, et al., "Genome Analysis of the Platypus Reveals Unique Signatures of Evolution," *Nature*, 8 May 2008, pp. 175-83. http://www.nature.com/nature/journal/v453/n7192/full/nature06936.html.

23. "L' ornitorinco sconfigge Darwin: Il o patrimonio genetico mette in crisi l' evoluzionism," *Corriere della Sera*, 11 May 2008, p. 33.

24. Ernst Mayr, *What Evolution Is*, p. 61.

25. Lewis Smith, "Oldest Platypus Fossil Identified," *The Sunday Times*, 22 January 2008. http://www.timesonline.co.uk/tol/news/world/us_and_americas/article3227763.ece. Retrived 9 April 2011.

26. "Platypus Thumbs Its Nose (or Bill) at Evolutionary Scientists," http://creationontheweb.com/index2.php?option=com_content&task=view&id=5783&pop=.

27. See Graham Lawton, "Cover Story: Uprooting Darwin's Tree," *NewScientist*, 24-30 January 2009, p. 39.

28. See, for example, Marlowe Hood, "Platypus Genome as Weird as Platypus," http://dsc.discovery.com/news/2008/05/07/platypus-genome-print.html; Massimo Pigliucci, "The Platypus, Evolution, and Why Piatelli-Palmarini Is Wrong," http://www.science20.com/rationally_speaking/the_platypus_evolution_and_why_piattelli_palmarini_is_wrong; and Elizabeth Finkel, "Platypus Genome Revealed," http://www.abc.net.au/rn/scienceshow/stories/2008/2240493. htm.

29. *The Origin of Species,* pp. 151-52.

30. Timothy Rowe, Thomas H. Rich, Patricia Vickers-Rich, Mark Springer, and Michael O. Woodburne, "The Oldest Platypus and Its Bearing on Divergence Timing of the Platypus and Echidna Clades," *Proceedings of the National Academy of Sciences, USA* 105, 29 January 2008, p. 1238.

31. Ibid, p. 1241.

32. *Evolution: A Theory in Crisis*, pp. 109-10.

33. *Science on Trial*, p. 197.

34. "This View of Life: Evolution's Erratic Pace," p. 14.

35. *What Evolution Is*, p. 14.
36. "Evolution," *Contemporary Botanical Thought*, ed. Anna M. MacLeod and L. S. Cobley (Edinburgh: Oliver and Boyd, 1961), p. 97.
37. Brian J. Alters and Sandra M. Alters, *Defending Evolution in the Classroom* (Toronto: Jones and Bartlett, 2001), p. 90.
38. *The God Delusion* (Boston: Houghton Mifflin, 2006), p. 79.

CHAPTER 8

THE "ORIGIN OF LIFE" DELUSION

More than 150 years after the publication of *Origin* (1859), the question of how species originate is still largely a mystery.[1]

—Leslie Alan Horvitz

The genome is not life itself. To understand what life is, we must make a radical switch of perception and view it at a variety of different levels, with interaction and feedback between gene, cell, organ, system, body, and environment.[2]

—Sir Patrick Bateson

A great deal of effort has been expended in finding theories (i.e., algorithms) for the origin of life without success. The reason may not be that we are not smart enough or that we have not worked hard enough. The reason may be that no structure or pattern exists which can be put into the terms of an algorithm of finite complexity . . . . It means that the *solution* to the problem is *undecidable*; it is beyond human reasoning.[3]

—Hubert P. Yockey

The production of varieties of plants (e.g., corn, grape fruit, roses, etc) and domestic animals (e.g., breeds of cats, dogs, horses, etc.) relies on the process of *artificial selection*. Darwin's experience with artificial selection in plant and animal breeding was instrumental in developing his theory on evolution by natural selection.

Human beings select the desirable traits in artificial selection, while presumably the natural environment does so in natural selection. If, according to evolutionists, natural selection replaces the role of God, then humanly controlled artificial selection should also qualify as another substitute for God's role. With artificial selection as a designing instrumentality that creates new species, in accordance with the evolutionary school of thought, human beings are central in the creation business. Therefore, in the laboratory scientists claim the creation of a new species of bacteria when, through treatment with antibiotics, a susceptible species changes into a more resistant species. In other words, new species of the same kind originate from preexisting species of that kind.

With artificial selection as a designing instrumentality two conclusions are noteworthy. First, species originate from preexisting life. This affirms the biological law of biogenesis. Species do not originate from non-living matter. Louis Pasteur empirically disproved the concept of spontaneous biogenesis. If we faithfully follow the scientific evidence and uphold the principle of uniformitarianism, then the origin of the first life on earth must be from preexisting life and not from inanimate matter. Since life cannot come from non-life, the hypothesis of spontaneous biogenesis or abiogenesis is an illusion! Second, to modify a living organism requires some form of intelligent guidance. Accordingly, a mindless process such as natural selection cannot engage in a productive selection process without the guidance of an external agency, since to act exclusively would imply that it is not mindless. As an illustration, when the digestive system sorts out food in the body, it does so not by chance but because it has been programmed to do so by a non-material command. Thus, in natural selection the same non-material programmer that is the architect of the genetic code presumably performs the role human beings play in artificial selection.

Darwin's theory of evolution stipulates that "species derive from other species by gradual evolutionary process and that the average level of each species is heightened by the 'survival of the fittest."[4] Since there are thousands of species, one would expect the title of his book to reflect this vastness by specifying "origins." However, in order to be more captivating and revolutionary, Darwin titled his work *The Origin of Species*, which gives the illusion of a common origin. Darwin's reference to a unique origin is found in the passage:

> Analogy would lead me one step further, namely to the belief that all animals and plants have descended from some one prototype. But analogy may be a deceitful guide. Nevertheless all living things have much in common, in their chemical composition, their germinal vesicles, their cellular structure, and their laws of growth and reproduction . . . . Therefore I should infer from analogy that probably all the organic beings which have ever lived on this earth have descended from some one primordial form, into which life was first breathed.[5]

However, this statement is too speculative to warrant his book's title *The Origin of Species*. The "origin" of an event implies its beginning or first existence. British physicist H. S. Lipson, in his brief review of *The Origin of Species*, accordingly writes, "Darwin's book—*On the Origin of Species*—I find quite unsatisfactory: it says nothing about the origin of *species*."[6] I fully concur. Natural selection does not create the products it selects; it selects products that have originated by another means. Therefore, we cannot relate natural selection to the origin of species. It is thus misleading to place natural selection in competition with the role of an Intelligent Creator. Natural selection cannot create a new product from scratch.

Analogy is a useful method for explaining things, but it becomes deceitful when used to arrive at a preconceived and dead-end conclusion. For instance, modern scientists are quick to conclude through analogy that, since the DNA of species are similar, they must have evolved from a common ancestor. They reject the alternative argument that such similarity could indicate a common designer. Their motivation is that, if we can explain all of life in material terms, then God is not necessary. Now it is common knowledge that the genetic program is more sophisticated than any developed by human intelligence. Summarized below are descriptions of DNA and cellular function from four sources.

The first description appears in *The Way Nature Works*:

> A single human cell contains 13 feet (4 metres) of DNA (deoxyribonucleic acid), packed into a nucleus less than a thousandth of an inch across. In this mass of tangled threads is contained all the information needed to make a human being. DNA directs development and maintains the life of an organism by instructing cells to make proteins—the versatile molecules on which all life depends. The cell's DNA is a vast library of coded commands: the long molecules are packaged into chromosomes on which genes are arranged like beads on string.[7]

Based on this account every reasonable person would imagine that intelligence was involved in the formation of DNA. If the information code needed to make a human being is assembled in a nucleus invisible to the naked eye, it suggests a designer of superior intellect. Human beings likewise, input information into computers to accomplish complex tasks. The programming does not come about by chance. Natural processes in this case cannot be deemed designing agencies.

Another description of DNA's complexity is offered by atheist Richard Dawkins:

> • The genetic code is not a binary code as in computers, nor an eight-level code as in some telephone systems, but a quaternary code, with four symbols . . . . DNA characters are copied with an accuracy that rivals anything modern engineers can do. They are copied

down the generations, with just enough occasional errors to introduce variety.[8]

- Each nucleus contains a digitally coded database larger, in information content, than all 30 volumes of *Encyclopaedia Britannica* put together. And this figure is for *each* cell, not all the cells of a body put together.[9]

Comparisons with computers, binary and quaternary codes, telephone systems, digital databases, and an encyclopaedia are indicators of abstract intelligence and hence a mind. If Dawkins' above assertions are not delusions, then God is not a delusion either.

In a third description Stephen Grocott of the University of Western Australia asserts:

> The complexity of the simplest imaginable living organism is mind-boggling. You need to have the cell wall, the energy system, a system of self-repair, a reproduction system, and means for taking in 'food' and expelling 'waste', a means for interpreting the complex genetic code and replicating it, etc., etc. The combined telecommunication systems of the world are far less complex, and yet no one believes they arose by chance.[10]

Grocott's account destroys the notion of chance in origin science and endorses the premise of supernatural intelligence.

Michael Denton's fascinating account, finally, in *Evolution: A Theory in Crisis* affirms Grocott's assertion:

- To grasp the reality of life as it has been revealed by molecular biology, we must magnify a cell a thousand million times until it is twenty kilometres in diameter and resembles a giant airship large enough to cover a great city like London or New York. What we would then see would be an object of unparalleled complexity and adaptive design.

On the surface of the cell we would see millions of openings, like the port holes of a vast space ship, opening and closing to allow a continual stream of materials to flow in and out. If we were to enter one of these openings we would find ourselves in a world of supreme technology and bewildering complexity. We would see endless highly organized corridors and conduits branching in every direction away from the perimeter of the cell, some leading to the central memory bank in the nucleus and others to assembly plants and processing units. The nucleus itself would be a vast spherical chamber more than a kilometer in diameter, resembling a geodesic dome inside of which we would see, all neatly stacked together in ordered arrays, the miles of coiled chains of the DNA molecules. A huge range of products and raw materials would shuttle along all the manifold conduits in a highly ordered fashion to and from all the various assembly plants in the outer regions of the cell.

- We would wonder at the level of control implicit in the movement of so many objects down so many seemingly endless conduits, all in perfect unison. We would see all around us, in every direction we looked, all sorts of robot-like machines. We would notice that the simplest of the functional components of the cell, the protein molecules, were astonishingly complex pieces of molecular machinery, each one consisting of about three thousand atoms arranged in highly organized 3-D spatial conformation. We would wonder even more as we watched the strangely purposeful activities of these weird molecular machines, particularly when we realized that, despite all our accumulated knowledge of physics and chemistry, the task of designing one such molecular machine—that is one single functional protein molecule—would be completely beyond our capacity at present and will probably not be achieved until at least the beginning of the next century. Yet the life of the cell depends on the integrated activities of thousands, certainly tens, and probably hundreds of thousands of different protein molecules.

- We would see that nearly every feature of our own advanced machines had its analogue in the cell: artificial languages and their decoding systems, memory banks for information storage and retrieval, elegant control systems regulating the automated assembly of parts and components, error fail-safe and proof-reading devices utilized for quality control, assembly processes involving the principle of prefabrication and modular construction. In fact, so deep would be the feeling of *deja-vu*, so persuasive the analogy, that much of the terminology we would use to describe this fascinating molecular reality would be borrowed from the world of late twentieth-century technology.

- What we would be witnessing would be an object resembling an immense automated factory, a factory larger than a city and carrying out almost as many unique functions as all the manufacturing activities of man on earth. However, it would be a factory which would have one capacity not equaled in any of our own most advanced machines, for it would be capable of replicating its entire structure within a matter of a few hours. To witness such an act at a magnification of one thousand million times would be an awe-inspiring spectacle.[11]

These are the testimonies of accomplished scientists about the awesomeness of the cell and DNA. For the unbiased these testimonies all point to a superior intelligence. Dawkins is a biologist, Grocott a chemist, and Denton a physician and molecular biologist. These men are scientists, not philosophers, and hence are not qualified to reach valid philosophical conclusions about God's existence. We thus require honest philosophers who follow the scientific evidence to draw a rational conclusion. In 2007 world-famous philosopher Antony Flew, after following these scientific findings as an atheist for half a century, concluded that God designed the world.

However, secularly minded scientists who seek only materialist explanations of the physical world disagree that divine intelligence is required. They ignore the fact that science cannot explain some of our natural experiences, such as the dynamics of the human mind, in materialist terms. However, some evolutionists are rational in their explanations. For instance, Robert Augros and George Stanciu write in their book *The New Biology*:

> What cause is responsible for the origin of the genetic code and directs it to produce animal and plant species? It cannot be matter because of itself matter has no inclination to these forms, any more than it has to the form of Poseidon or to the form of a microchip or any other artifact. There must be a cause apart from matter that is able to shape and direct matter. Is there anything in our experience like this? Yes, there is: our own minds. The statue's form originates in the mind of the artist, who then subsequently shapes matter, in the appropriate way. The artist's mind is the ultimate cause of that form existing in matter, even if he or she invents a machine to manufacture the statues. For the same reasons there must be a mind that directs and shapes matter in organic forms. Even if it does so by creating chemical mechanisms to carry out the task with autonomy, this artist will be the ultimate cause of those forms existing in matter. This artist is God, and nature is God's handiwork.[12]

It sometimes happens that more faith is required to deny the truth than humbly to accept it. Common sense dictates that it is naïve to think that matter can explain everything in this universe. Only the spiritually handicapped will deny the coexistence of a physical and spiritual world. Small wonder, then, the Scriptures insist that it is the spiritually foolish who deny the existence of God.

For those who follow the scientific evidence, the genetic-program analogy points to superior intelligence. For evolutionists analogy is a useful guide in deducing that similarity in DNA presupposes common ancestry. However, for the *origin* of genetic information and complexity, which points to God's existence, analogy is no longer adequate. Consistent with Feynman's principle for scientific integrity, Lipson studies both worldviews and contends that creation is the only reasonable answer to the question of our origin. He admits that he does not personally like this conclusion but that scientific evidence must override philosophical preference. Lipson asserts that many people, not necessarily creationists, have doubts about the evolutionist worldview.[13] And it will remain that way because the doubts outweigh compelling pieces of scientific evidence.

Modern scientists set a poor precedent when they opted to favour the hypothesis of abiogenesis against the scientific law of biogenesis that presupposes a creator. If scientists, who are supposed to show professionalism by following the evidence wherever it leads, pick and choose only data that favour their preconceptions, they mislead the public. Nobel laureate Ernst Boris Chain is right in his comment: "The view that scientists are objective, dispassionate, impartial and tolerant is a myth . . . . Their power of logical thinking is also not above that of other professions."[14] Chain is not being sarcastic; the subject of evolution proves his point. How can anyone imagine that mindless, blind, and dumb natural processes can create abstract programs as blueprints for life?

The public must understand why evolutionism cannot survive without embracing the unscientific hypothesis of abiogenesis. Christopher Carlisle and W. Thomas Smith, Jr. explain in *The Complete Idiot's Guide to Understanding Intelligent Design*:

> If abiogenesis is possible, then the way is made clear for a strictly scientific explanation of life. Human beings can be seen as a natural product of material existence, with no need for intervention by any transcendent, "intelligent" agent . . . . The chance origination of life from inorganic matter allows the unguided, material process that indeed is evolution.[15]

Knowing full well that to them as professionals preference (abiogenesis) should not contradict fact (biogenesis), evolutionists rush to declare evolution by natural selection a scientific fact. What is the reason behind this rush to (mis)judgment? It is that abiogenesis removes God from the picture. Carlisle and Smith further explain:

- Remember that Darwin had insisted on natural selection being self-sufficient—that is, requiring no transcendent, guiding hand to direct life to some predetermined end. Many neo-Darwinists take Darwin's insistence to mean more than that God isn't required. By asserting that all of life can be materially explained, they assert that God does not exist.[16]
- If Darwin took God out of the equation in unprecedented ways, it was his adherents who came to perceive evolution as final proof there is no God.[17]

Evolutionists' inability to establish scientifically the validity of their belief in abiogenesis places them in the same boat as the advocates of intelligent design. Creationists, by faith, believe that God created life (biogenesis); evolutionists, by faith, believe that life sprang spontaneously from dead matter (abiogenesis). Experiential knowledge and empirical science, however, are consistent only with biogenesis. The onus is on evolutionists to establish abiogenesis, which until now they have been unable to accomplish. The evolution paradigm thus remains a myth. You cannot build a tree of life with fictitious branches and no roots and argue that it is real. When an investigation is grounded on fictitious ideas such as abiogenesis, any conclusion can be tailored to fit.

NASIM distinguishes between science and religion as follows:

In science, explanations *must* be based on evidence drawn from examining the natural world. Scientifically based observations or experiments that conflict with an explanation eventually *must* lead to modification or even abandonment of that explanation. Religious faith, in contrast, does not depend only on empirical evidence, is not necessarily modified in the face of conflicting evidence, and typically involves supernatural forces or entities.[18]

Today, with virtually no knowledge of how life originated and despite empirical evidence that dismisses any likelihood of life's spontaneous emergence from inanimate matter, the neo-Darwinian theory based on evolution by natural selection has been neither modified nor abandoned. This suggests that the theory is predicated solely on pseudoscientific/religious grounds. Such a verdict agrees with Lipson's remark that evolution (more accurately evolutionism) has become in a sense a religion.[19] Michael Ruse, Professor of Philosophy at Florida State University, supports Lipson's claim by admitting that "evolution is indeed a secular religion—a full-fledged alternative to Christianity."[20]

Science should not compete with philosophy. If, according to NASIM, "Science is a way of knowing that differs from other ways in its dependence on empirical evidence and testable explanations,"[21] then modern scientists have reneged by advocating evolutionism. Today the integrity of science as a discipline is compromised because many of its practitioners give precedence to materialist explanations rather than objective truth. One is compelled to ask, "Without the theory of evolution by natural selection, would the world be very different from what it is today?" Academically and technologically the answer is no; socially, politically, and religiously the answer is yes.

In its effort to promote the evolutionist worldview, NASIM highlights science's achievements over the ages as follows:

Every day we rely on technologies made possible through the application of scientific knowledge and processes. The computers and cell phones which we use, the cars and airplanes in which we travel, the medicines that we take, and many of the foods that we eat were developed in part through insights obtained from scientific research. Science has boosted living standards, has enabled humans to travel into Earth's orbit and to the Moon, and has given us new ways of thinking about ourselves and the universe.[22]

Except for "new ways of thinking about ourselves and the universe," however, these are achievements without the stamp of evolution on them. Moreover, all the pioneers in science behind these achievements—Isaac Newton, Charles Babbage, James Clerk Maxwell, Johannes Kepler, Louis Pasteur, Joseph Lister, James Young Simpson, and Alexander Fleming—were creationists.

What these pioneers thought about the origin of the universe or species had no methodological bearing on their practice of science. Why then are these parameters so crucial to modern scientific investigation? The scientific community of yesteryear never hashed out in courts of law what science is, primarily because it separated scientific practice from philosophical preference. Professional integrity demands that scientists follow the evidence wherever it leads, notwithstanding preferred philosophical conclusions.

THE SCIENTIFIC MYTH OF ABIOGENESIS AND THE ORIGIN-OF-LIFE SCIENCE FOUNDATION

Modern scientists sinned against nature when they chose to believe in spontaneous generation or abiogenesis, which amounts to an effort to remove God from the equation of life. Without any empirical justification, evolution in effect became a pseudoscientific religion. To try to recover from this misstep, the discipline's community established the Origin-of-Life Science Foundation to explain how life can emerge from non-life. Distancing itself from "creation science" or "intelligent design," as previously discussed, the Foundation encourages the pursuit of natural-process explanations and mechanisms within nature itself. In order to stimulate global interest, an annuity designated as "The Origin-of-Life Prize" has been set aside to award the winner/s for "proposing a highly plausible mechanism for the spontaneous rise of genetic instructions in nature sufficient to give rise to life."[23] Establishment of both the Origin-of-Life Science Foundation and its "Origin-of-Life Prize" confirm two things. First is the fact that evolutionism as a religion is not different from creationism since both posit faithful adherence to preconceived beliefs. Second, evolutionism is big business.

The Origin-of-Life Science Foundation's objectives are spelled out in the following proclamation:

> No theory of genetic information is complete without a model of mechanism for the *source* of such prescriptive information within Nature. It is not sufficient for a submission to the Prize to limit discussion of prescriptive information (instruction) theory to its replication, transmission, modification, or matrix of information retention. *All submissions must address the source of the prescriptive information through non-supernaturalistic natural processes.* Which of the four known forces of physics (strong and weak nuclear forces, electromagnetic force, and gravity), or what combination of these forces, produced *prescriptive, functional* information, and how? What is the empirical evidence for this kind of prescriptive information (instruction) spontaneously arising within Nature?[24] (Emphasis theirs.)

The above quotation is evidence that evolutionism has no scientific foundation. Scientists are just as ignorant about evolutionism as they are of creationism. And confirming its degree of ignorance the Foundation wants to be enlightened on the following points:

- The initial writing of this prescriptive information by nature, not just the modification of existing genetic instruction through mutation.
- The ability of that information (instruction) not only to give the directions or orders of what should be

done, but to bring to pass those orders in the form of actual physical molecules, products, and life processes.[25]

By requesting the "initial writing of this prescriptive information by nature," the Foundation stepped beyond the proper domain of science. The Foundation provides definitions for terms such as "theory," "mechanism," "prescriptive information," and "genetic code," but it offers no definition for "nature."[26] What is the difference, one might ask, between "nature" and God?

It is puzzling that we here are pondering extremely intelligent questions about events that some scientists believe happened spontaneously by random chance and mindless natural processes. With the wrong premise scientists cannot come up with the right answers, but the Origin-of-Life Science Foundation demands mechanistic answers to how "seemingly unintelligent natural processes could have written such highly prescriptive recipe/ message linguistic-like code."[27]

Evolutionists argue that absence of fossil evidence is not evidence of absence in defending their worldview. However, biophysicist Hubert P. Yockey, a scientist with remarkable integrity, cautiously describes the answer to the question of species' origin as "unknowable" and "beyond human reasoning."[28] Since God is unknowable through science, Yockey's conclusion is consistent with the Scriptures, where God declares Himself to be both the "beginning" and the "end." Investigations into the origin and end of life are, therefore, beyond the boundaries of science.

Based on information theory, Yockey argues: "The same genetic code, the same DNA, the same amino acids and genetic message that unites all organisms, independent of morphology, proves that the theory of evolution is well established as any in science."[29] Because Yockey also asserts that "The origin of life, like the origin of the universe, is unknowable,"[30] he cannot issue any conclusive statement on the merits of Darwin's theory on the origin of species. The most he can say is that "The Darwinian view that all the organic beings which have ever lived on this earth have descended from some one primordial form," and even that claim is unsubstantiated. With no hard fossil or laboratory evidence of organisms evolving from simple to complex, the appropriate message from information theorists should be, "The origin of species, as the origin of genetic code, life, or the universe, is unknowable." Yockey frames his interpretations of the evidence to promote Darwin's theory of evolution as he tries to distance himself from creationists to pacify evolutionists. This is so because his evidence that the origin of life is unsolvable as a scientific problem places him on the fence between creationism and evolutionism. Yockey, however, makes it very clear that he rejects creationism and Intelligent Design.[31]

We characterize organisms not only by similarities but by dissimilarities as well. How does information theory account for the dissimilarity of organisms? More specifically, how can information theory explain the genetic programs that made the platypus partly reptilian, partly avian, and partly mammalian? Is this a special case of the segregated, linear, and digital message in the genome? Yockey must understand that, no matter how vital the data from information theory is, it can only complement or supplement and not override or supplant fossil data.

Richard Dawkins asks in *The God Delusion*, "But if science cannot answer some ultimate question, what makes anybody think that religion can?"[32] The answer lies in the domain of religion. Religion may not tell us "how," as NASIM requires, but it can tell us who created the universe. When accomplished scientists disagree on philosophical grounds, we must look outside their community for guidance. Unbiased and distinguished philosophers must step in to lead the public to a correct conclusion. This is what philosopher Antony Flew did when he declared there is a God after fifty years of professing atheism.

EVOLUTION: MYTH OF A UNIFYING THEORY

The ambition of modern biologists in defense of evolution theory is transparent. In their book *Defending Evolution*, Brain J. Alters and Sandra M. Alters write:

Why then should science students learn about evolution?
A simple answer is that evolution is the basic context
of all the biological sciences. Take away this context,
and all that is left is disparate facts without the thread
that ties them all together. Put another way, evolution
is the explanatory framework, the unifying theory.
It is indispensable to the study of biology, just as
the atomic theory is indispensable to the study of
chemistry.[33]

However, is a unifying theory in biology possible? While the
laws of physics or chemistry are universal, the laws in biology
can only be a broad generalization. Nobel laureate Francis Crick
explains the reasons as follows:

- There is really nothing in biology that corresponds to special
  and general relativity, or quantum electrodynamics, or even such
  simple conservation laws as those of Newtonian mechanics: the
  conservation of energy, of momentum, and of angular momentum.
  Biology has its "laws," such as those of Mendelian genetics, but
  they are often only rather broad generalizations, with significant
  exceptions to them. The laws of physics, it is believed, are the
  same everywhere in the universe. This is unlikely to be true of
  biology.
- Theorists in biology should realize that it is extremely unlikely
  that they will produce a useful theory (as opposed to a mere
  demonstration) just by having a bright idea distantly related to
  what they imagine to be facts.[34]

The desire for a unifying theory is, therefore, not a valid reason
for students to learn evolution. Discussing the same point, Andre
Brown writes:

A single *meaningful* Theory of Biology is not possible.
I don't discount the possibility of more powerful
theories of many *aspects* of biology such as increasingly
broad and quantitative "laws of evolution," or the

eventual quantitative understanding of a cell, but a single theory, expressed as a usefully small number of equations or a usefully short algorithm, that accounts for biology from the level of molecules to populations will never be.[35]

The discussion of evolution as a unifying theory would not have been necessary had studies been limited to bacteria-to-bacteria transformations. Evolutionists' objective, however, is to unite bacteria-to-bacteria and bacteria-to-human transitions under the same mechanisms so that the latter can be declared a scientific fact on the strength of the former. The assumption that the same mechanism is responsible for the two autonomous fields of evolution is false until proven. A unifying theory must address how one aspect can lead to the other.

The only report in this area that I have come across is that of Daniel R. Brooks and E. O. Wiley. Their book titled *Evolution as Entropy: Toward a Unified Theory of Biology* maintains that "instructional information resides in organisms, comprises a physical array, and is 'closed.'"[36] Viewing evolution as the result of proximal and ultimate causes, they argue that neo-Darwinism based on the three major principles (selection, competition, and dispersal) is a relatively complete theory of proximal causes in evolution. They also contend that, by unjustifiably extending proximal causes to the level of ultimate causes, neo-Darwinians produce an incomplete and relatively weak theory.[37] Brooks and Wiley's theory is radically different from Danny Vendramini's "Unified Teem Theory of Evolution, Perception, Emotions, Behaviour and Inheritance."[38] Vendramini proposes a second evolutionary process that regulates the inheritance of environmental information.

Simply put, we need to develop a theory that accounts for extraevolution (macroevolution) and shows how intraevolution (microevolution) processes can be deduced therefrom. Biologists must overcome the illusion that successive micro events will result in macro events. Abrupt gaps in the fossil record suggest that natural biological systems are not developed by either continuous arithmetical progression (from simple to complex form) or regression (from complex to simple form). In physics, for instance, the speed of light limits the validity of Newton's theory on motion. In biology the harmful and sometimes lethal nature of large mutation limits the validity of the theory of evolution. To account for the failure of Newton's theory at the cosmological level (high speed), Einstein developed a new theory of motion. Einstein, knowing full well that Newton's theory was valid at the terrestrial level (low speed), worked on a more general or unifying theory that was not based on extrapolation from what was already known to be true. Instead, he developed a relativistic formula which not only fits the results at the cosmological level but also reduces to the nonrelativistic formula at the terrestrial level.

The current theory of evolution is adequate only for small mutations in consistent agreement with microevolutionary phenomena. As the Newtonian mechanics is a special case of Einstein's theory, microevolution should be a special case of macroevolution. A fully unifying theory of evolution, in light of the above discussion, is one that would adequately account for the unknown (macroevolution) and from which the known (microevolution) could be deduced as a special case. It *cannot* be what we have now—a theory from which the unknown (macroevolution) is extrapolated from the known (microevolution). Since mutations are more harmful than beneficial, a fully unifying theory is impossible if mutations are the raw materials for evolution. The limiting effect of mutations is attested by abrupt gaps in the fossil record. In other words, a valid unifying theory must address those abrupt gaps. The boundaries they define are real and not imposed by chance. Based on information theory, there is a temptation to ignore the relevance of observed gaps. Information theory must explain gaps in the fossil record and not simply ignore them.

On Cynthia Yockey's website,[39] for instance, appears the following chart by Hubert Yockey regarding "Intelligent Design" versus "scientific reality":

False assertions by believers in Intelligent Design:	The scientific reality:
Gaps in the fossil record constitute valid objections to Darwin's theory of evolution because they are spaces for the miraculous appearance of species that have not evolved from any other source.	The fundamental consideration in evolution is the genome, not the fossil record. Gaps in the fossil record do not matter. What matters is that there are no gaps in the continuity of the genome from the origin of life to the present. It is the continuity of the genome that shows the connectedness of all life—living, extinct, and yet-to-be-evolved. That means there are no gaps in which species miraculously appear, as Intelligent Design falsely claims.

Gaps in the fossil record are well-defined facts and, therefore, a scientific reality. Information about the genome cannot contradict fossil data. Gaps in the fossil record matter because they reflect gaps in the living world. Creationists and evolutionists alike know that the gaps are an essential part of fossil data. These gaps contradict an evolutionary worldview, according to which we expect continuity in the fossil record. Continuity of knowledge about the genome is inconclusive in that the genome is only part of the puzzle. Here the problem has to do with starting assumptions. In reaching his conclusion, Yockey assumes the similarity of genomes as evidence of a common ancestor. If, for the sake of scientific integrity, Yockey also assumes that similarity is possible evidence of a common designer, then, since the model based on a common designer will account for dissimilarities among organisms his conclusions will change, and the gaps revealed by fossils will be explained.

Yockey asserts that the origin of DNA is undecidable.[40] We know that in problem-solving, once the beginning of a problem is unknowable, one cannot proceed to a reliable conclusion. Yockey's conclusion that "there is no need for records of intermediate structures or a fossil record because there are no gaps in the genome" is, therefore, based on the wrong premise.[41] Yockey also asserts that the message in the genome is segregated, linear, and digital. If these properties are not adequate to account for gaps in the fossil record, he should consider other points of view such as the nature of continuity. According to Yockey,

> The genome, which is the non-material information programmed into DNA, has definite starting and stopping points the information transcribed from it. For example, the genome for making a mouse does not run forever—it stops when it has made a mouse.[42]

Based on this evidence a more appropriate description of the relationship among species would be "separated piecewise linearly" as opposed to "connected piecewise linearly," for this presumably would explain the abrupt gaps observed in fossil data. If the genome evolves through a Markov chain, the programmer presumably limits this process to changes within the domain of plant and animal breeding. This explains why bacteria-to-human evolution cannot be achieved scientifically. Yockey's goal is to link evolution to the genome in order to exonerate Darwinian theory. It does not matter to him what the fossil data actually reveal.

Scientists such as Sir Isaac Newton, Albert Einstein, and George Wald believed that a superior mind governs our universe. To these empiricists the function of a genetic code would have been further evidence for the existence of an intelligent mind behind this universe. Scientists who think otherwise must prove their honesty by providing evidence of where non-intelligence has generated coded information that governs chemical reaction. Yockey, for instance, should explain the data that compel him to decide that "segregated, linear, and digital" DNA structure attests to a common ancestry and, hence, confirms the theory of evolution.

Evolutionism advocates materialism, which is why evolutionists "cannot allow a Divine Foot in the door."[43] If according to information theory the genome, which is non-material, is the engine of evolution, it brings the evolutionary paradigm into question. An organism has (1) a genome, which is the non-material information programmed into DNA (a material substance) and which is measurable in terms of bits and bytes; and (2) a mind, which is also non-material but cannot be measured in bits and bytes. The question then is, "Does mind evolve?" Why or why not, and how? The larger point is that similarity in DNA or genome is a trivial argument for either evolution or creation. One cannot conclude whether an organism evolved or not because its genome can be measured. The genome is unique mainly in separating living organisms from non-living matter. Genome measurements cannot adequately distinguish between living organisms. Therefore, measurement of an organism's genome is only part of an intricate problem.

The genetic code (the non-material map) does not evolve.[44] Based upon partial information, Yockey concludes that existence of "the same genetic code, similar DNA, the same amino acids" proves that the theory of evolution is as well established as any in science. In reaching this conclusion, he ignores the "mindset" of species. Programmed information comes from a mind. There is no evidence to suggest that the greater the similarity of genomes, the more closely related the mindsets of species are. Would Yockey honestly conclude that existence of "the same genetic code, similar DNA, dissimilar mindset, the same amino acids" proves that the theory of evolution is as well established as any in science? Hence Yockey does not have complete evidence to rule in favour of either evolution or creation, since information theory does not account for all the non-material components of organisms.

Yockey argues that "Evolution and the origin of life are two separate problems,"[45] which of course is true. My point, however, is that the origin of species and the origin of life are two related problems. If the origin of life cannot be explained, the origin of species also cannot be explained. In defense of Darwin, Yockey writes:

Darwin never proposed a theory for the origin of life in his scientific publications. What he wrote about the origin of life in Chapter XV of the sixth edition of *The Origin of Species* (1872) is as follows, "It is no valid objection that science as yet throws no light on the far higher problem of the essence or origin of life."[46]

Darwin is misleading here. The origin of species and the origin of life are each unsolvable as scientific problems, a fact that explains why Darwin's book was all about evolution and natural selection with no rigorous scientific discussion of the origin of species. Since life could not have evolved spontaneously, Yockey maintains that the origin of life should be accepted as an axiom of biology.[47] My argument is that, once the origin of life is accepted as an axiom of biology, science automatically loses its mandate to address the origin of species through the Darwinian theory of evolution by natural selection.

According to Yockey, an axiom is "an elementary fact that cannot be proved or derived from any other facts and therefore must be taken as a staring point."[48] The problem Yockey faces in his bailout of evolutionism is that the "axiom argument" can go either way. In his description of DNA, Yockey states:

> The messages in the DNA sequences are similar to programs used by modern computers. mRNA acts like the reading head on a Turing machine, which moves along the DNA sequence to read off the genetic message to the proteome.[49]

Since computers are created by intelligent minds, given Yockey's analogy, DNA represents the elementary fact of creation by an intelligent mind. Because this fact cannot be proved or demonstrated scientifically, it must be accepted as an axiom of science. Thus, as a starting point, DNA as the blueprint for life originated from an intelligent mind. And since God is the only Being who has claimed credit for creating the world, we can logically deduce that God created DNA.

Taking all these facts into consideration, creation by God is a better explanation of the "unknowables" in science, as it justifies both the realities of the fossil record (the document of the cosmological history of species) and the genome (the message or non-material information programmed in DNA). Any theory, whether a unifying theory or information theory, for biological systems must reflect the abrupt gaps in fossil data and in the living world in order for it to be conclusive and authentic.

In sum, contrary to Yockey's view, an intelligent designer is needed to explain the genome, DNA, and the genetic code. The gaps in morphology or the fossil record are valid objections to the Darwinian theory of evolution as long as similar gaps are evident in the living world. Life is immaterial, and the answer to the origin of life lies outside the domain of materialism. The diversity of life on planet Earth cannot be explained by the simple arithmetic of bits and bytes in the genomes of organisms. The fact that God created life and DNA should be accepted as an axiom of reliable science. I conclude with the following words of David Berlinski:

> We do not know how the universe began. We do not know why it is there. Charles Darwin talked speculatively of life emerging from a "warm little pond." The pond is gone. We have little idea of how life emerged, and cannot with assurance say that it did. We cannot reconcile our understanding of the human mind with any trivial theory about the manner in which the brain functions. Beyond the trivial, we have no other theories. We can say nothing of interest about the human soul. We do not know what impels us to right conduct or where the form of the good is found . . . . While science has nothing of value to say on the great and aching questions of life, death, love, and meaning, what the religious traditions of mankind *have* said forms a coherent body of thought.[50]

Because evolutionists are unable to demonstrate empirically the various stages of the bacteria-to-human or atoms-to-traits postulate, God's claim to have created the world is the only honest answer to our quandary.

NOTES

1. *The Complete Idiot's Guide to Evolution* (Indianapolis: Alpha Books, 2002), p. 174.
2. "What is Life?" qtd. in Denise Noble, *The Music of Life: Biology Beyond the Genome* (Oxford: Oxford University Press, 2006), backcover of dust-jacket.
3. *Information Theory: Evolution and the Origin of Life* (Cambridge: Cambridge University Press, 2005), p.188.
4. *The Origin of Species* (New York: Signet, 2003), p. 1.
5. *The Origin of Species* (New York: Penguin, 1968), p. 455.
6. "Origin of Species," *NewScientist*, 90 (1981), p. 452.
7. Jill Bailey, ed., *The Way Nature Works* (New York: MacMillan, 1997), p. 96.
8. *River Out of Eden: A Darwinian View of Life* (London: Weidenfeld and Nicolson, 1996), pp. 17, 19.
9. *The Blind Watchmaker* (London: Penguin, 2006), pp. 17-18.
10. "Inorganic Chemistry," *In Six Days: Why 50 Scientists Choose to Believe in Creation*, ed. John F Ashton, (Sydney: New Holland Publishers, 1999), p. 136.
11. *Evolution: A Theory in Crisis* (Bethesda, MD: Adler and Adler, 1986), pp. 328-29.
12. *The New Biology: Discovering the Wisdom in Nature* (Boston: New Science Library, 1987), p. 191.
13. "Origin of Species," p. 452.
14. "Social Responsibility and the Scientist in Modern Western Society," *Perspectives in Biology and Medicine* 14.3 (1971), p. 357.
15. *The Complete Idiot's Guide to Understanding Intelligent.Design* (New York: Alpha Books, 2006), p. 133.
16. Ibid, p. 180.
17. Ibid, p. 174.
18. *Science, Evolution, and Creationism* (Washington, D.C.: National Academies Press, 2008), p. 12.
19. "A Physicist Looks at Evolution," *Physics Bulletin* 31 (May 1980), p. 138.
20. "How Evolution Became a Religion: Creationists Correct?" *National Post*, 13 May 2000, National Edition, pp. B1, B3.

21. *Science, Evolution, and Creationism*, p. 12.

22. Ibid, p. xi.

23. The Origin-of-Life Prize. Gene Emergence Project. Description and Purpose of the Prize. http://www.us.net/life/rul_desc.htm.

24. The Origin-of-Life Prize. Discussion. http://www.us.net/life/rul_disc.htm.

25. Ibid.

26. The Origin-of-Life Prize. Definitions. http://www.us.net/life/rul_defi.htm.

27. Ibid.

28. *Information Theory, Evolution, and the Origin of Life* (Cambridge: Cambridge University Press, 2005), p. 188.

29. *Information Theory, Evolution, and the Origin of Life*, p. 181.

30. Ibid, p. 184.

31. See "Hubert Yockey: Reply to FTE Amicus Brief." http://ca.search.yahoo.com/search?p=Hubert+Yockey+reply+to+FTE+amicus+brief& fr=yfp-t-501&toggle=1&cop=&ei=UTF-8. Retrieved 12 November 2008.

32. *The God Delusion* (Boston: Houghton Mifflin, 2006), p. 56.

33. *Defending Evolution: A Guide to the Creation/Evolution Controversy* (London: Jones and Bartlett, 2001), p. 104.

34. *What Mad Pursuit: A Personal View of Scientific Discovery* (New York: Basic Books, 1988), pp. 138, 142.

35. "A Unified Theory of Biology?" http://biocurious.com/a-unified-theory-of-biology. Retrieved 7 July 2008.

36. *Evolution as Entropy: Toward a Unified Theory of Biology* (Chicago: University of Chicago Press, 1988), p. 36.

37. Ibid, pp. 367-68.

38. "The Second Evolution: The Unified Teem Theory of Evolution, Perception, Emotions, Behaviour and Inheritance." http://thesecondevolution.com/ Introduction.pdf.

39. "Scientific Reality vs. Intelligent Design's False Claims. The Problem Is Getting Caught in Behe's Tar Baby, Not Darwin's Black Box." http://www.cynthiayockey.com/pages/1/index.htm. Retrieved 12 November 2008.

40. *Information Theory, Evolution, and the Origin of Life*, p. 188.

41. "Scientific Reality vs. Intelligent Design's False Claims."

42. "Hubert Yockey: Reply to FTE Amicus Brief."

43. Richard Lewontin, "Billions and Billions of Demons," *The New York Review*, 9 January 1997, p. 31.

44. "Scientific Reality vs. Intelligent Design's False Claims."

45. "Hubert Yockey: Reply to FTE Amicus Brief."

46. "Scientific Reality vs. Intelligent Design's False Claims."

47. "Hubert Yockey: Reply to FTE Amicus Brief."

48. Ibid.

49. "Scientific Reality vs. Intelligent Design's False Claims."

50. *The Devil's Delusion: Atheism and Its Scientific Pretensions* (New York: Crown Forum, 2008), pp. xiii-xiv.

CHAPTER 9

GOD IS NO DELUSION BUT SCIENTISM IS

Scientists of the stature of Carl Sagan, Stephen Jay Gould, E. O. Wilson, and Richard Dawkins, just to name a few, offer the theory of evolution to us already wrapped up in the alternative "faith" of scientific materialism.[1]

—John F. Haught

Scientism is the belief that there is no truth outside of what can be demonstrated by the natural sciences . . . . When presented with facts that seem to contradict Darwin's scenario, the disciples of scientism often behave like devout religionists whose faith has been challenged. Some get belligerent; others become cagey. Still others have been known to blurt out, "There is no God; therefore it *had* to be that way." This, however, is not science but "scientism" of a very crude sort.[2]

—George Sim Johnston

I do not understand how the scientific approach alone, as separated from a religious approach, can explain an origin of all things . . . . In my view, the question of origin seems always left unanswered if we explore from a scientific view alone. Thus, I believe there is a need for some religious or metaphysical explanation if we are to have one.[3]

—Nobel laureate Charles H. Townes

Professor John F. Haught provides the following definition of "scientism":

- Scientism may be defined as "the *belief* that science is the only reliable guide to truth." Scientism, it must be emphasized, is by no means the same thing as science. For while science is a modest, reliable, and fruitful method of learning some important things about the universe, *scientism* is the assumption that science is the *only* appropriate way to arrive at the totality of truth . . . . Devotees of scientism place their trust in scientific method itself, but no more than religious believers can they scientifically demonstrate the truth of this faith. They trust deeply in the power of science to clear up all confusion about the world, but they cannot scientifically justify this trust without logical circularity . . . . Scientism is in fact no less a conflation of science with belief than is "scientific creationism."
- And scientism . . . is closely associated with a correlative set of beliefs known as "scientific materialism." Scientific materialism is a belief system built on the assumption that all reality, including life and mind, is reducible to and completely explainable in terms of lifeless matter . . . . In philosophical jargon, scientism is the epistemological component and materialism the metaphysical ingredient of an influential modern creed that functions for many

scientists in very much the same way that religion functions for its devotees.[4]

Human beings as products of a process cannot choose their manufacturer/s simply by rule of philosophical preference in science. Their manufacturer must claim them. That is precisely what the Judaeo-Christian God Yahweh did through enactment of the Creation Sabbath Law in the Decalogue. Belief in a supernatural Being cannot be deemed a delusion if there is a documented eyewitness account of its intervention in world history and interaction with historical figures. God's intervention in ancient Egyptian and Israeli history and specific dealings with Pharaoh Ahmose to free ancient Israelites from slavery and Moses to claim credit for having created the universe, is well documented. Moses, in his review of these events with the Arabian Desert generation of Israelites, cautions:

> Remember the day you stood before the LORD your God at Horeb . . . . You came near and stood at the foot of the mountain . . . . Then the LORD spoke to you out of the fire . . . . He declared to you his covenant, the Ten Commandments, which he commanded you to follow and then wrote them on two stone tablets. And the LORD directed me at that time to teach you the decrees and laws you are to follow in the land that you are crossing the Jordan to possess. (Deut. 4: 9-37 NIV)

Here, we are not dealing with religious belief or faith but with concrete historical evidence of supernatural intervention. This summary of an eyewitness account of historical events cannot be deemed a delusion.

In this chapter, countering Richard Dawkins' erroneous assumptions, I intend to argue that belief in God is not a delusion. I will support my arguments by drawing on Antony Flew's realistic account in *There is a God*. I then will conclude with a personal testimony for the existence of God.

RICHARD DAWKINS' *THE GOD DELUSION*

The Historical Perspective

Clinton Richard Dawkins was born into an Anglican family. At the early age of nine years he started doubting the existence of God but subsequently reconverted. He then reverted to atheism after studying Darwin's theory of evolution by natural selection. Dawkins believes that learning Darwinism should ultimately lead to atheism, despite his own broad knowledge of biblical history. Historical facts, unlike scientific facts, are not subject to change. For this reason Dawkins' objective is to create doubt in the minds of his readers about the authenticity and credibility of these accounts. I will illustrate with two examples.

First, we consider the story about Jephthah, Gilead's head of state and military commander, which Dawkins narrates as follows:

> In Judges, chapter 11, the military leader Jephthah made a bargain with God that, if God would guarantee Jephthah's victory over the Ammonites, Jephthah would, without fail, sacrifice as a burnt offering 'whatsoever cometh forth of the doors of my house to meet me, when I return'. Jephthah did indeed defeat the Ammonites ('with a very great slaughter', as is par for the course in the book of Judges) and he returned home victorious. Not surprisingly, his daughter, his only child, came out of the house to greet him (with timbrels and dances) and—alas—she was the first living thing to do so. Understandably Jephthah rent his clothes, but there was nothing he could do about it. God was obviously looking forward to the promised burnt offering, and in the circumstances the daughter very decently agreed to be sacrificed. She asked only that she should be allowed to go into the mountains for two months to bewail her virginity. At the end of this time she meekly returned, and Jephthah cooked her. God did not see fit to intervene on this occasion.[5]

Despite Dawkins' slanted rhetoric, note that the Scriptures do not say Jephthah "cooked" his daughter but that he "did to her as he had vowed." Vows to God that involve a sacrifice are usually limited to animals (Lev. 27:9-13). Moreover, sacrifices of sons and daughters constitute an abomination to God, as is evident in the following quotation:

> The LORD your God will cut off before you the nations you are about to invade and dispossess. But when you have driven them out and settled in their land, and after they have been destroyed before you, be careful not to be ensnared by inquiring about their gods, saying, "How do these nations serve their gods? We will do the same." *You must not worship the LORD your God in their way, because in worshiping their gods, they do all kinds of detestable things the LORD hates. They even burn their sons and daughters in the fire as sacrifices to their gods.* (Deut. 12:29-31; emphasis mine)

Obviously it is not God's will that Jephthah's daughter be sacrificed as a vow fulfillment. God's position regarding the sacrifice of human beings was not declared in words or document at the time of Abraham. In the case of Jephthah, God's will regarding sacrifices are stipulated in God's law, and there was no need for God to intervene. God trusted Jephthah to do the right thing under the given circumstances. What did Jephthah do? It is very unlikely that he sacrificed his daughter, as Dawkins would have us believe. She bewailed her virginity, not her untimely death.

As a leader Jephthah would have known that vows were limited to offerings of animals, houses, land, and other personal effects (Lev. 27:9-25). Let us give him the benefit of doubt that he was not naïve. Moreover, no priest would undertake to conduct human sacrifice, which was an abomination to God. The only human sacrifice God accepted was in the person of Jesus Christ. If Jephthah's daughter were sacrificed, as some scholars erroneously claim, her death would undermine the unique significance of God's atonement for sin through Christ as the sacrificial lamb. Would God embrace an abomination in exchange for a vow?

Dawkins' ulterior motive is to convince his readers that there is no God. Why is he unable to say, "On Mount Sinai, before an audience of Israelites, God is reported to have claimed credit for creating life and the universe"? Dawkins' argument against God's existence would be honest if he acknowledged this historical event and then proceeded to discredit the story or the claim. Why is Dawkins silent on this point? Because to present the fact that God publicly claimed credit for creation would prove God's existence and put to rest the creationism-evolutionism controversy.

Second, we consider the story about God's enforcement of the Creation Sabbath Law. In this law God, in speech and in print, claims credit for having created the world in six days. To commemorate this event, the Sabbath day was declared sacred. Capital punishment at that time was the verdict for deliberate violation of the Creation Sabbath Law. According to Numbers 15:32-36, a man apparently profaned this law and was punished. Dawkins uses this incident as evidence against God in Chapter 7 of *The God Delusion*.

In this biblical account one thing is clear: because the Sabbath breaker undermined God's claim on creation and ownership of the universe, he was subject to divine justice. The Israelites at the time were reluctant to enforce the death penalty, but God sanctioned their doing so. So serious was the Sabbath violation that God cited it as the major reason for scattering the Israelites across the globe. Disbelief in claims issued directly by God is tantamount to the sin of atheism. God enforced the terms of the covenant to affirm that His cosmological claims were not mythical.

Dawkins' antipathy toward God does not destroy His credentials as Creator. Dawkins' disgruntled epithets for God—"a petty, unjust, unforgiving control-freak; a vindictive, bloodthirsty ethnic cleanser; a misogynistic, homophobic, racist, infanticidal, genocidal, filicidal, pestilential, megalomaniacal, sadomasochistic, capriciously malevolent bully"[6]—do not alter the truth of God's supremacy as

Creator. No amount of such petulant and childish rebellion can obscure the truth. Nonetheless, the power of Dawkins' rhetoric should not be dismissed lightly. Every aspect of his challenge to theism must be addressed.

The Imaginary Perspective

Unable to provide concrete scientific evidence to support Darwin's theory of evolution by natural selection, Dawkins resorts to hand-waving probability arguments to woo the media and the public. In *The God Delusion* he thus writes:

- Any entity capable of intelligently designing something as improbable as a Dutchman's Pipe (or a universe) would have to be even more improbable than a Dutchman's Pipe.
- What is it that makes natural selection succeed as a solution to the problem of improbability, where chance and design both fail at the starting point? The answer is that natural selection is a cumulative process, which breaks the problem of improbability up into small pieces. Each of the small pieces is slightly improbable, but not prohibitively so. When large numbers of these slightly improbable events are stacked up in series, the end product of the accumulation is very, very improbable indeed, improbable enough to be far beyond the reach of chance.
- [A] God capable of designing a universe, or anything else, would have to be complex and statistically improbable.
- [A] God who is capable of sending intelligible signals to millions of people simultaneously, and of receiving messages from all of them simultaneously, cannot be simple.[7]

Dawkins also tries to convince the public that Einstein espoused similar views:

Let me sum up Einsteinian religion in one more quotation from Einstein himself: 'To sense that behind anything that can be experienced there is a something that our mind cannot grasp and whose beauty and sublimity reach us only indirectly and as a feeble reflection, this is religiousness. In this sense I am religious.' In this sense I [Dawkins] too am religious, with the reservation that 'cannot grasp' does not have to mean 'forever ungraspable'. But I prefer not to call myself religious because it is misleading. It is destructively misleading because, for the vast majority of people, 'religion' implies 'supernatural'.[8]

From Dawkins' point of view the phrase "'cannot grasp'" aligns Einstein and his own way of thinking. However, let us consider another quotation from Einstein that discusses religion but in which the phrase "cannot grasp" is absent:

My religion consists of a humble admiration of the illimitable superior spirit who reveals himself in the slight details we are able to perceive with our frail and feeble minds. That deeply emotional conviction of the presence of a superior reasoning power, which is revealed in the incomprehensible universe, forms my idea of God.[9]

Here Einstein maintains that we are able to "perceive" God as an "illimitable superior spirit." On this basis Einstein, unlike Dawkins, publicly proclaims that he is not an atheist. In the humble spirit of scientific inquiry, Einstein, conceives of God as a quantum that cannot be fully grasped. In the arrogant spirit of scientism, Dawkins believes that nothing is beyond the human "grasp."

We next consider Dawkins' argument that "[A] God capable of designing a universe, or anything else, would have to be complex." Thomas Crean in *God Is No Delusion* illustrates brilliantly that "a thought cannot be a material thing, nor can it be caused by a material thing, nor can it be the property of a material thing." Crean concludes, therefore, that "thought" is independent of matter and is something spiritual.[10] Discounting Dawkins' views on complexity, Crean writes:

> [I]n the realm of thought, greater simplicity is a mark of greater perfection. The better the knower, the simpler his manner of knowing. Far from supposing, then, that a being perfect enough to know and design the entire universe must be extremely complex, we ought to suppose that he would be extremely simple. Nor should it bother us if we cannot imagine what his knowledge would be 'like'. A dog, whose knowledge is limited to what his senses can perceive, could not imagine how any being could have 'a million oak trees' as a *single* object of knowledge. For him, an oak tree is something that he looks at or sniffs. Since he cannot sniff a million trees simultaneously, he cannot imagine how this number of trees could be known by a single act. We, who possess conceptual knowledge as well as sense knowledge, do understand how this is possible. But if we desired by means of our own experience to grasp the nature of divine knowledge, we should be in the position of a dog trying by its own experience of the world to understand its master's thoughts.
>
> God knows all things in a single act by knowing Himself. All things bear some resemblance to God: they are limited 'reflections' of His unlimited nature. Therefore, by knowing Himself He also knows all creatures that could ever exist. His act of knowledge is thus entirely simple. He does not have one thought about the stars, another about plants, another about mankind. God is one eternal 'thought', by which He knows both Himself and all possible and actual creatures.[11]

Accordingly, what is highly complex in the material world may be simple in the non-material realm. Building on this argument, Crean remarks:

> When we speak of scientific laws, we are emphasizing that not everything in the world happens by chance, and that things have a predetermination to act in definite ways, like sodium reacting with water, even before they do act. Yet scientific laws by themselves, of course, cannot do anything. They cannot cause any event to occur. Only things can cause events to occur. And when the thing that acts is, like sodium, incapable of having intentions, some other thing must have assigned its action to it, to explain why it produces one action rather than another, or rather than none at all.[12]

Crean thus establishes simple concepts about the material world that differ markedly from Dawkins' materialistic school of thought. What appears complex to us in the material world is speculatively quite simple in the non-material world. We shall next attach a scientific meaning to Crean's theological deduction.

On the question of the evolution of complex things, Dawkins in *The Blind Watchmaker,* quotes the findings of Peter Atkins as follows:

> He [Atkins] asks what the minimum necessary physical conditions are, what is the minimum amount of design work that a very lazy Creator would have to do, in order to see to it that the universe and, later, elephants and other complex things would one day come into existence. The answer, from his point of view as a physical scientist, is that the Creator could be infinitely lazy. The fundamental original units that we need to postulate, in order to understand the coming into existence of everything, either consist of literally nothing (according to some physicists), or (according to other physicists) they are units of the utmost simplicity, far too simple to need anything so grand as deliberate Creation.[13]

Scientist Atkins examines the material world and concludes that the fundamental units we require to understand how things came to be are simple and reflect a Creator who is "infinitely lazy." Theologian Crean analyzes the non-material world and postulates that we require an extremely simple Creator. Infinite simplicity is an indicator of infinite knowledge, per the philosophical principle of Occam's razor.

What is responsible for the timeless preservation of the genes that Dawkins assumes to account for our existence? According to Deepak Chopra, fragile molecules like DNA should have disintegrated long ago under the pressure of entropy or random mistakes through mutation, heat, wind, sunlight, and radiation. He attributes DNA's persistence to an invisible source outside space and time, a field that can create something new and then remember it. This explanation makes more sense than the mindless-selection arguments that Dawkins offers. Chopra concludes: "Dawkins falls prey, not to the delusion of God, but to the delusion of an almighty chance acting mindlessly through matter. He cannot admit the possibility of an ordering force in Nature."[14]

Below is an excerpt from Gary Parker's *Creation: Facts of Life*, which describes a scene in which Dawkins is asked to give an example that justifies his scientific theory:

- In a video production featuring several evolutionist and creationist leaders and skeptics, Dawkins argued eloquently that millions of years of mutation and natural selection would serve as a "blind watchmaker," producing all appearance of design among living things without any help from some supernatural Designer. Then, in a quiet, non-threatening voice, not knowing what the answer would be, the narrator asked Dawkins to give an example of a mutation that *adds information.*
- The usually effusive Dawkins gestured, opened his mouth, but stopped before he spoke. With his eyes shifting back and forth as if searching for some answer, he started to speak several times, but always checked himself. Finally, after a long embarrassing silence, the program resumed with Dawkins speaking on a

different subject—leaving unanswered the ultimate question, the origin of genetic information.

- Yet, molecules-to-man evolution is all about phenomenal expansion of genetic information. *It would take thousands of information-adding mutations to change "simple cells" into invertebrates, vertebrates, and mankind.* If there were any scientific merit at all to mutation-selection as a mechanism for evolution, Dawkins' reply should have been enthusiastic and overwhelming. "My three favorite examples of mutations adding information are . . . . Excellent examples among plants are . . . among insects are . . . among bacteria are . . . ." His answer, instead, was silence, and with no mechanism to add genetic information, the "evolutionary tree" can't grow.[15]

Because the evolutionary tree cannot grow, some evolutionists are uprooting it. Wise move!

The late physicist Richard Feynman, a staunch advocate for scientific integrity, was right: "The first principle of scientific thought that corresponds to a kind of utter honesty is that you must not fool yourself—and you are the easiest person to fool."[16] He reminds scholars that "the idea is to try to give all the information to help others to judge the value of their contribution, not just the information that leads to judgment in one particular direction or another."[17] Had Dawkins followed the scientific evidence instead of his materialistic preference, he would not be so compromised an apologist. After the embarrassing incident above, in order to save face, Dawkins responded as follows:

> In September 1997, I allowed an Australian film crew into my house in Oxford without realizing that their purpose was creationist propaganda. In the course of a suspiciously amateurish interview, they issued truculent challenge to me to "give an example of a genetic mutation or an evolutionary process which can be seen to increase the information in the genome." It is the kind of question

only a creationist would ask in that way, and it was at this point I tumbled to the fact that I had been duped into granting an interview to creationists—a thing I normally don't do, for good reasons . . . . I shall try to redress the matter now in constructive fashion by answering the original question, the "Information Challenge", at adequate length—the sort of length you can achieve in a proper article.[18]

Dawkins then proceeded to give an elaborate answer, contending that the information content of a biological system is another name for its complexity and indicating that he has addressed this subject in three of his books—*The Blind Watchmaker*, *River Out of Eden*, and *Climbing Mount Improbable*. Elsewhere he develops the idea that natural selection gathers up information from the environment and builds it into the genome.

Can mutation via natural selection, however, write new genetic programs? This is apparently in line with one of the questions the Origin-of-Life Science Foundation wants to know and for the answer to which, as discussed earlier, it is offering a million-dollar prize. Dawkins' accusation that only creationists would pose this kind of question is, therefore, untrue. Anybody who is meticulous about correct scientific information would ask such a question! The bad news for Dawkins, though, is that none of his books is on the list of the Foundation's seventy suggested texts. The Foundation needs real science, not fantasy or delusionary stories. Understanding how a genetic program works is one thing, but explaining its creation is a very different matter. According to information theorist Herbert Yockey, this is something forever ungraspable.[19] Let us not be fooled, then, by the disciples of scientism. If the origin of DNA and, hence, life is unknowable, science should remain silent since it cannot credit either intelligence or non-intelligence. To choose between the two is to opt for either theism or atheism, and this is not the province of science.

One would expect that Dawkins in *The God Delusion* would address this crucial question about how mutation-natural selection *"adds information"* in the genome, but he simply recycles his usual hyperbolic rhetoric: "Evolution by natural selection produces an excellent simulacrum of design, mounting prodigious heights of complexity and elegance."[20] Instead of explaining how this happens with concrete examples to back up his claim, Dawkins sidesteps the issue by writing, "You need to be steeped in natural selection, immersed in it, swim about in it, before you can truly appreciate its power."[21] This is not an acceptable explanation of how we allegedly have evolved.

What other possible explanations should the public know? Gary Parker was once a strong advocate of evolutionism, and so he understands Dawkins' frustration:

> The problem with evolution is not some shortcoming in Dawkins, however. The problem is with the fundamental nature of information itself. The information in a book, for example, cannot be reduced to, nor derived from, the properties of the ink and paper used to write it. Similarly, the information in the genetic code cannot be reduced to, nor derived from, the properties of matter nor the mistakes of mutations; its message and meaning originated instead in the mind of its Maker . . . . Although mutations may corrupt it and selection may sort variations into different environments, it was not a "blind watchmaker" that composed the genetic script for each kind of organism, but a Creator with a plan and purpose and eyes wide open.[22]

Dawkins rejects the notion of a Creator in stating, "[A]ny God capable of intelligently designing something as complex as the DNA/protein replicating machine must have been at least as complex and organized as that machine itself."[23] Is natural selection, the designing instrumentality in the evolutionist worldview, as complex and organized as the DNA/protein replicating machine? However, Dawkins still insists: "To explain the origin of the DNA/protein machine by invoking a supernatural Designer is to explain precisely nothing, for it leaves unexplained the origin of the designer."[24] Similarly, to explain the origin of the DNA/protein machine by invoking natural selection is to explain precisely nothing, for it leaves unexplained the origin of natural selection.

Dawkins wraps his narratives in subjective improbability discussions and occasionally uses a handful of computer simulations to explain the evolutionist worldview. In *The God Delusion*, however, he also uses Scripture to discredit the probability of God's existence. It thus would be worthwhile to use the same evidence here to rebut his arguments.

The probability of the Judaeo-Christian God's influence on world history is essentially the same as that of Moses or the Pharaohs. According to the Abrahamic Scriptures, God spoke to Moses in the manner of a person speaking to a friend (Exod. 33:11; Deut. 34:10). The confirming evidence is provided in Exodus 34 and Numbers 12:

> (a) The LORD told Moses, "Prepare two stone tablets like the first ones. I will write on them the same words that were on the tablets you smashed. Be ready in the morning to come up Mount Sinai and present yourself to me there on the top of the mountain." . . . So Moses cut two tablets of stone like the first ones. Early in the morning he climbed Mount Sinai as the Lord has told him, carrying the two stone tablets in his hands. (Exod. 34:1-2, 4 NLT)

> The Lord again wrote the terms of the covenant—the Ten Commandments—on them and gave them to me. They were the same words the Lord had spoken to you from the heart of the fire on the mountain as you were assembled below . . . . I stayed on the mountain in the Lord's presence for forty days and nights, as I had done the first time. (Deut. 10:4, 10 NLT)

> When Moses came down the mountain carrying the stone tablets inscribed with the terms of the covenant, *he wasn't aware that his face glowed because he had spoken to the LORD face to face.* And when Aaron and the people of Israel saw the radiance of Moses' face, they were afraid to come near him. (Exod. 34:29-30 NLT; (Emphasis mine)

(b) While they were at Hazeroth, Miriam and Aaron criticized Moses because he had married a Cushite woman. They said, "Has the Lord spoken only through Moses? Hasn't he spoken through us, too?" . . . So immediately the LORD called to Moses, Aaron, and Miriam and said, "Go out to the Tabernacle, all three of you!" And the three of them went out. Then the LORD descended in the pillar of cloud and stood at the entrance of the Tabernacle. "Aaron and Miriam!" he called, and they stepped forward. And the LORD said to them, "Now listen to me! Even with prophets I the LORD communicate by vision and dreams. But that is not how I communicate with my servant Moses. He is entrusted with my entire house. *I speak to him face to face, directly and not in riddles!* He sees the LORD as he is. Should you not be afraid to criticize him?" (Num. 12:1-8 NLT; emphasis mine)

Here is the record of God's interaction with Pharaoh through Moses:

Then the Lord said to Moses, "You are to say everything I command you, and your brother Aaron is to tell Pharaoh to let the Israelites go out of his country. But I will harden Pharaoh's heart, and though I multiply my miraculous signs and wonders in Egypt, he will not listen to you. Then I will lay my hand on Egypt and with mighty acts of judgement I will bring out my divisions, my people the Israelites. And the Egyptians will know that I am the Lord, when I stretch out My hand against Egypt and bring out the sons of Israel from their midst." Moses and Aaron did just as the Lord commanded them . . . . Moses was eighty years old and Aaron eighty-three when they spoke to Pharaoh. (Exod. 7:1-5, 7 NIV)

These biblical passages are provided to rebut Dawkins' thesis "Why There Almost Certainly Is No God."[25]

Another measure of the reality of the Judaeo-Christian God can be deduced in terms of the probability of the historical existence of the wilderness generation of ancient Israelites. To them God verbally and in print claims credit for having created the universe (Exod. 19-20; Deut. 5). This is the only written record in world history in which, through an awesome display of power, a supernatural being has claimed credit as the Creator before a nation. Historical events, fortunately, are not like pseudoscientific speculations that can be changed by wishful thinking. The Judaeo-Christian God is, therefore, not "vanishingly probable" as Dawkins claims. For his views to prevail, Dawkins has to prove beyond a reasonable doubt that these events are myths. So far he has been unable to do so.

Further evidence that discounts Dawkins' argument about God is in the person of Jesus Christ. The prophet Isaiah spoke of Christ's relationship with God as follows:

- The virgin will be with child and will give birth to a son, and will call him Immanuel (God with us). (Isa. 7:14 NIV)
- For to us a child is born, to us a son is given, and the government will be on his shoulders. And he will be called Wonderful Counselor, Mighty God, Everlasting Father, Prince of Peace. (Isa. 9: 6 NIV)

Therefore, with reference to world history, the probability of God's reality is as good as that of the historical existence of Jesus Christ or Alexander the Great.

Marketing Atheism

Alister McGrath, Oxford University Professor of Historical Theology, writes:

When I read *The God Delusion* I was both saddened and troubled. How, I wondered, could such a gifted popularizer of the natural sciences, who once had such a passionate concern for the objective analysis of evidence, turn into such an aggressive antireligious propagandist with an apparent disregard for evidence that was not favorable to his case? Why were the natural sciences being so abused in an attempt to advance atheist fundamentalism? I have no adequate explanation. Like so many of my atheist friends, I simply cannot understand the astonishing hostility that he [Richard Dawkins] displays toward religion.[26]

McGrath holds an Oxford D. Phil. in molecular biophysics and an Oxford Doctorate of Divinity. As an ex-atheist, therefore, McGrath has impeccable credentials to comment on Dawkins' abuse of both science and religion in his effort to promote atheism under the banner of scientism. It also be helpful to view Dawkins from a philosophical standpoint, since he mixes science with philosophy to contradict the existence of God.

For fifty years Antony Flew championed atheism from a philosophical perspective; Dawkins, on the other hand, marketed atheism in popularizing evolutionism. The years 2006 and 2007 are noteworthy for two conflicting books from these renowned exegetes. After Dawkins published *The God Delusion* in 2006, Flew released *There Is a God: How the World's Most Notorious Atheist Changed His Mind*. After several decades of fellowship with the disciples of scientism, Flew decided to follow the scientific evidence rather than accept the opinions of materialists. So, while Flew was freeing himself from the spell of scientism, Dawkins was ascending the global pulpit of atheism. Dawkins arrives with an aggressive marketing strategy that targets both the academic world and the family unit.

In the academic world Dawkins uses the yardstick of "Einsteinian religion" to determine which scientists are serious about God. He quotes studies that show an inverse relationship between intelligence (or level of education or interest in science) and belief in God. Dawkins thus writes:

Great scientists who profess religion become harder to find through the twentieth century, but they are not particularly rare. I suspect that most of the more recent

ones are religious only in the Einsteinian sense which,
I argued in Chapter 1, is a misuse of the word.[27]

Dawkins then discusses studies that enable him to conclude that
"The overwhelming majority of FRS [Fellows of the Royal Society],
like the overwhelming majority of US academicians, are atheists."[28]
Such atheist membership is presumably one of Dawkins' major
reasons for believing that God is a delusion. However, according
to McGrath, most of the unbelieving scientists whom he has come
across "are atheists on grounds other than their science; they bring
those assumptions *to* their science rather than basing them *on*
their science."[29] For instance, Kate Hammer, in her article on the
2011 Templeton prize laureate Martin Rees, wrote:

> Martin Rees, a 68-year-old Cambridge astrophysicists,
> has made essential contributions to science's understanding
> of black holes and the origins of the universe. He's
> an atheist, but one whose ideas about the lucky
> conditions of the universe leave room for an all-powerful
> deity . . . .
>
> In one of his books, *Just Six Numbers*, Dr. Rees
> argued that the perfect tuning was neither a mere accident
> nor the act of a benign creator. Instead, he said, "an
> infinity of other universes may well exist" where the
> constants are set differently. Some would be too sterile
> to support life, others too short-lived. Ours happens to
> be just right . . . .
>
> Dr. Rees has been criticized by atheistic hardliners
> such as Richard Dawkins, who called him a "compliant
> Quisling" for being a "believer in belief."
>
> Though neither scientist subscribes to religion, Dr.
> Rees doesn't go so far as to mock religion, like Dr.
> Dawkins. Instead, in books and lectures, Dr. Rees has
> let his scientific ideas spill over into philosophy.[30]

Most atheists, scientists among them, rest their case on the evils
of organized religion and the world's imperfections. However, as

Flew remarks in *There Is a God*, "The excesses and atrocities of organized religion have no bearing whatsoever on the existence of God, just as the threat of nuclear proliferation has no bearing on the question of whether $E = mc^2$."[31]

Science cannot prove whether God exists or not; hence, the number of unbelieving scientists is of no significance. People should not be swayed by Dawkins' misguided ideas about God since his objective is to market atheism at all cost. In *The Dawkins Delusion?* McGrath writes about his encounter with a young man after a lecture in which he gave a point-by-point, evidence-based rebuttal to Dawkins' views on religion:

> Since the publication of my book *Dawkins' God* in 2004, I am regularly asked to speak on its themes throughout the world . . . . After one such lecture, I was confronted by a very angry young man. The lecture had not been particularly remarkable. I had simply demonstrated, by rigorous use of scientific, historical and philosophical arguments, that Dawkins' intellectual case against God didn't stand up to critical examination. But this man was angry—in fact, I would say he was furious. Why? Because, he told me, wagging his finger agitatedly at me, I had "destroyed his faith." His atheism rested on the authority of Richard Dawkins, and I had totally undermined his faith. He would have to go away and rethink everything. How *dare* I do such a thing![32]

If adults can be lured into atheism so easily, how much more so would children without any previous religious upbringing?

Dawkins' declared position is to expose children to no religious views on the origin of the cosmos so that in science classes a child can be molded to fit the pattern of an "Evolution Child." According to Dawkins, knowledge of evolution has religious implications. He contends:

> Any creationist lawyer who got me on the stand could instantly win over the jury simply by asking me: 'Has your

knowledge of evolution influenced you in the direction
of becoming an atheist?' I would have to answer yes,
and at one stroke I would have lost the jury.[33]

If Dawkins believes that teaching evolution leads to atheism
and advocates a science curriculum that teaches only evolution,
is he not guilty of child abuse? Does it not occur to him that the
only way *not* to abuse children, as he conceives the situation, is
not to teach them anything about their origins at any stage of their
educational development? When students are trained to accept
evolutionism as a scientific fact, the expectation is for them to
view creationism as false. Are the tenets of evolutionism that posit
"survival of the fittest" or "human life without purpose or meaning"
better lesson than the religious doctrine of "love for all"? When life
is presented as having no intrinsic value, its preservation becomes
meaningless. Little wonder that, with evolutionism entrenched
in our modern science curricula, physical safety in educational
institutions can no longer be guaranteed.

In presenting evolutionism as a scientific fact, teachers become
de facto recruiters for atheism. Dawkins himself takes pleasure in
fulfilling this goal, as is evident in his testimony about converting
a creationist student who had graduated from a fundamentalist
college in the United States. Dawkins' reward for this achievement
was a T-shirt with the inscription, 'Evolution—The Greatest Show
on Earth—The Only Game in Town!'[34] How fulfilling it must have
been for Dawkins to see this Christian young man transformed
into an "Evolution Child"!

ANTONY FLEW'S *THERE IS A GOD*

Antony Flew, the only son of a Methodist minister, was once
the world's most famous atheist. Flew's conversion, prompted by
scientific discoveries and consolidated by philosophical arguments,
suggests that most people are atheists for reasons other than
science itself. It takes humility, open-mindedness, intellectual
integrity, and courage for the world's leading atheist to abandon
his faith, especially in the wake of today's so-called "new atheism."

HarperCollins, the publisher of Flew's historic book *There Is a God*, recounts:

> A wave of modern atheists have taken center stage and brought the long-standing debate about the existence of God back into the headlines. Spearheaded by Richard Dawkins, Sam Harris, and Christopher Hitchens, this "new atheism" has found a power place in today's culture wars. Although this movement has been billed as "new," the foundation of its argument is indebted to philosopher Antony Flew and his groundbreaking paper "Theology and Falsification," the most widely reprinted philosophical publication of the last half century. Flew built his highly acclaimed academic career publicly debunking the existence of God. But now the renowned philosopher has arrived at the opposite conclusion and officially joined the other side. In *There Is a* God, Flew discloses his newfound belief in a God who created the universe. Flew earned his fame by arguing that one should presuppose atheism until evidence of a God surfaces. He now believes that such evidence exists. *There Is a God* reveals for the first time the scientific discoveries and philosophical arguments that turned him from a staunch atheist into a believer. With refreshing openness to argument and an absence of the anger and hostility that have been hallmarks of the "new atheism," Flew shows how his commitment to following the argument wherever it leads resulted, to his own astonishment, in his conversion to belief in a creator God. Certain to be read and discussed for years to come, *There Is a God* will forever change the debate about the existence of God.[35]

In his "conversion," however, Flew encountered considerable opposition from "the other side." Here is his own account:

> I had previously written that there was room for a new argument to design in explaining the first emergence

of living from nonliving matter—especially where this first living matter already possessed the capacity to reproduce itself genetically. I maintained that there was no satisfactory naturalistic explanation for such a phenomenon. These statements provoked an outcry from critics who claimed that I was not familiar with the latest work in abiogenesis. Richard Dawkins claimed that I was appealing to a "God of the gaps."[36]

I am not surprised at Dawkins' attempt to discourage Flew from endorsing theism. Confident that the scientific facts were in his favour, Flew continues:

It is true that protobiologists do have theories of the evolution of the first living matter, but they are dealing with a different category of problem. They are dealing with the interaction of chemicals, whereas our questions have to do with how something can be intrinsically purpose-driven and how matter can be managed by symbol processing. But even at their own level, the protobiologists are still a long way from any definitive conclusions.[37]

Flew had become convinced that biologists do not have all the answers to the metaphysical question of life's origin.

As the biblical saying goes, "The truth shall set you free." Flew was delivered from scientism and atheism, which both Newton and Einstein declared nonsensical. Flew indicates that modern science highlights three aspects of nature that point to God:

The first is the fact that nature obeys laws. The second is the dimension of life, of intelligently organized and purpose-driven beings, which arose from matter. The third is the very existence of nature. But it is not science alone that has guided me. I have also been helped by a renewed study of the classical philosophical arguments.[38]

Flew, however, insists that he found God via reason and not on the grounds of religious faith:

> I must stress that my discovery of the Divine has proceeded on a purely natural level, without any reference to supernatural phenomena. It has been an exercise in what is traditionally called natural theology. It has had no connection with any of the revealed religions. Nor do I claim to have had any personal experience of God or any experience that may be called supernatural or miraculous. In short, my discovery of the Divine has been a pilgrimage of reason and not of faith.[39]

Flew's conversion, therefore, was not based on religious conviction but on the same scientific evidence that the disciples of scientism and materialism have inappropriately credited as verifying evolution by natural selection.

Predictably, there were furious responses to Flew's conclusion, but he stood his ground by clarifying his change in conviction:

> You might ask how I, a philosopher, could speak to issues treated by scientists. The best way to answer this is with another question. Are we engaging in science or philosophy here? When you study the interaction of two physical bodies, for instance, two subatomic particles, you are engaged in science. When you ask how it is that those subatomic particles—or anything physical—could exist and why, you are engaged in philosophy. When you draw philosophical conclusions from scientific data, then you are thinking as a philosopher.[40]

Flew continues:

> Of course, scientists are just as free to think as philosophers as anyone else. And, of course, not all scientists will agree with my particular interpretation of the facts they generate. But their disagreements will have to stand on their own

two philosophical feet. In other words, if they are engaged in philosophical analysis, neither their authority nor their expertise as scientists is of any relevance. This should be easy to see. If they present their views on the economics of science, such as making claims about the number of jobs created by science and technology, they will have to make their case in the court of economic analysis. Likewise, a scientist who speaks as a philosopher will have to furnish a philosophical case. As Albert Einstein himself said, "The man of science is a poor philosopher."[41]

Flew's explanation is consistent with John Phillip's view that science and religion address different questions: "Science asks what and how, while religion asks why."[42] The scientific evidence points to God, but leading scientists seem obsessed with scanting it. On the matter of the origin of the cosmos, only a rigorous philosopher like ex-atheist Flew can heroically follow the evidence to wherever it leads. Michael Ruse is another philosopher who bolted from the scientist camp to argue that Darwin's molecules-to-human paradigm of evolution by natural selection is a full-fledged religion and a spurious substitute for the teachings of Christianity.

To assert his intellectual authority, Flew eloquently critiques both Darwin's and Dawkins' views on natural selection as a misleading interpretation of natural processes:

> *On Darwin*: In my book *Darwinian Evolution,* I pointed out that natural selection does not positively produce anything. It only eliminates, or tends to eliminate, whatever is not competitive. A variation does not need to bestow any actual competitive advantage in order to avoid elimination; it is sufficient that it does not burden its owner with any competitive disadvantage . . . . Darwin's mistake in drawing too positive an inference with his suggestion that natural selection produces something was perhaps due to his employment of the expressions "natural selection" or "survival of the fittest" rather than his own ultimately preferred alternative, "natural preservation."

On Dawkins: I went on to remark that Richard Dawkins'
The Selfish Gene was a major exercise in popular
mystification. As an atheist philosopher, I considered this
work of popularization as destructive in its own ways as
either *The Naked Ape* or *The Human Zoo* by Desmond
Morris . . . . Genes, of course, can be neither selfish nor
unselfish any more than they or any other nonconscious
entities can engage in competition or make selections . . . .
But this did not stop Dawkins from proclaiming that his
book "is not science fiction; it is science . . . . We are
survival machines—robot vehicles blindly programmed
to preserve the selfish molecules known as genes." . . .
He added, sensationally, that "the argument of this book
is that we, and all other animals, are machines created
by our genes." If any of this were true, it would be no
use to go on, as Dawkins does, to preach: "Let us try
to teach generosity and altruism, because we are born
selfish." No eloquence can move programmed robots.
But in fact none of it is true—or even faintly sensible.
Genes, as we have seen, do not and cannot necessitate
our conduct. Nor are they capable of the calculation
and understanding required to plot a course of either
ruthless selfishness or sacrificial compassion.[43]

The role of a selection process is to weed out undesired features,
but Darwinian evolution requires natural selection to develop new
features and genetic information, and that is just not possible.
Natural selection is a mindless process, yet Dawkins believes
that it generates complexities. To attribute the same evidence to
God, Dawkins argues that God must be exceedingly complex in
order to generate any degree of complexity. Flew challenges the
rationale of Dawkins views as follows:

This strikes me as a bizarre thing to say about the concept
of an omnipotent spiritual Being. What is complex about
the idea of an omnipotent and omniscient Spirit, an idea
so simple that it is understood by all adherents of the

three great monotheistic religions—Judaism, Christianity, and Islam? Commenting on Dawkins, Alvin Plantinga recently pointed out that, by Dawkins's own definition, God is simple—not complex—because God is spirit, not a material object, and hence does not have parts.[44]

A simple God generates what to humanity seems complex.

Science and hence biology is the study of matter, and that is why scientists advocate materialism as the only explanation of nature. However, the information in DNA is non-material; therefore, its origin is not a scientific problem. Accordingly, scientists, and in particular biologists, are not the appropriate professionals to provide valid philosophical conclusion on the origin of life. Flew as a philosopher legitimately points out the sloppy approach of the scientific enterprise to origin-of-life studies:

> Most studies on the origin of life are carried out by scientists who rarely attend to the philosophical dimension to their findings. Philosophers, on the other hand, have said little on the nature and origin of life. The philosophical question that has not been answered in origin-of-life studies is this: How can a universe of mindless matter produce beings with intrinsic ends, self-replication capabilities, and "coded chemistry"? Here we are not dealing with biology, but an entirely different category of problem.[45]

The answer to the philosophical question that Flew specifies cannot be provided by biological science and this corroborates Jerry Fodor and Massimo Piattelli-Palmarini's assertion that the story about the evolution of phenotypes belongs not to biology but to history.[46] There is only one rational conclusion that scientists and philosophers alike can reach. On the ultimate question of "How do we account for the origin of life?" Flew bases his conclusion on the view of Nobel laureate George Wald:

> The Nobel Prize-winning physiologist George Wald once famously argued that "we choose to believe the

impossible: that life arose spontaneously by chance." In later years he concluded that a preexisting mind, which he posits as the matrix of physical reality, composed a physical universe that breeds life.[47]

"This, too," adds Flew, "is my conclusion. The only satisfactory explanation for the origin of such 'end-directed, self-replicating' life as we see on earth is an infinitely intelligent Mind."[48] Wald and Flew's views of our origin is inconsistent with Darwin and Dawkins' theory of evolution by natural selection. Who is right and who is wrong here?

In 1992, responding to H. D. Lewis's essay on "The Existence of the Universe as a Pointer to the Existence of God," Flew said:

> I do not see how anything within our universe can be either known or reasonably conjectured to be pointing to some transcendent reality behind, above, or beyond. So why not take the existence of that universe and its most fundamental features as the explanatory ultimates?[49]

Here we notice Flew speaking from the pulpit of scientism and atheism. Between 1992 and 2007, however, he changed his mind and endorsed theism.

Flew's stance, buttressed by Wald, is solid for two reasons. First, it comes from a distinguished scientist who changed his mind from belief in abiogenesis (preferred scientific myth) to biogenesis (established scientific fact). Second, it comes from a world-renowned philosopher who switched from atheism to theism because of overwhelming scientific evidence. His reconsidered outlook is consistent with that of Einstein, who declared:

> Certain it is that a conviction, akin to religious feeling, of the rationality or intelligibility of the world lies behind all scientific work of a higher order . . . . This firm belief, a belief bound up with deep feeling, in a superior mind that reveals itself in the world of experience, represents my conception of God.[50]

This is the God Isaac Newton deemed as personal and credited for all his breakthrough scientific discoveries.

MICHAEL EBIFEGHA'S *THE DARWINIAN DELUSION*

Initially I resisted the idea of including any personal or experiential testimony in this book. After glancing through Dawkins' account of "The Argument from Personal Experience" in *The God Delusion*, however, I became convinced that I should make an exception.

An integral part of my communication with God is from the Scriptures through words/instructions that were issued to and documented by the people who were chosen by God in the ancient world. Since God is immutable, one would expect Him to adopt the same pattern of communication in any generation. People have asked me to provide evidence of God's presence in my life. Answers such as one's faith in God through Jesus Christ are usually not the ones they fancy. Instead, they want some material evidence of one's faith in God. So the evidence I give as proof of God's existence are testimonies of: (1) supernatural intervention in response to my faith in God; and (2) divine intervention through others to address my concerns. I must stress that there are other ways by which people can experience God's presence in their lives.

Dawkins asserts that arguments from personal experience are generally weak.[51] He ridicules testimonies that posit personal visions or auditory impressions of God. Dawkins goes to great lengths to describe a childhood experience in which wind gusting through a keyhole created sounds that, for the impressionable, could pass for a ghost's voice. My account of personal experience will therefore be limited to God's interaction with me through others.

My father Godfrey Allison Ebifegha was so committed to the service of God that he established the Holy Ghost Devotees Church (HGDC) in Nigeria. HGDC is classified as a "spiritual Church," which proclaims that "God is Spirit and must be worshipped in spirit and in truth." In his secular life my father worked for the Ministry of Mines and Power, the nature of his job taking him from one location to another. To ensure that the transfers did not compromise our standard of education, my uncle Kostom Ebifegha raised us at our hometown of Oyobu. We completed elementary and part of secondary education under the guidance of this uncle. Prior to the establishment of HGDC, we adopted the pagan customs of our uncle. Upon the establishment of HGDC, we relinquished the pegan customs, but our uncle remained a pagan.

One of the doctrinal tenets of HGDC at its inception was a belief in divine healing. Common physical disorders at the time were wounds, particularly on the feet. Such wounds were rarely healed without treatment at a nearby clinic or dispensary.

When my cousin Lawrence and I, both under fifteen years, developed these wounds on our feet, we decided, in order to justify our faith in divine healing, against an injection by the local hygienist. So we simply applied hot water and olive oil as a lubricant to the infected areas. For over a month there was no improvement, and we started to notice signs of decay. As our condition continued to deteriorate, my pagan uncle intervened. I still declined to receive any medical attention, but he persuaded Lawrence and took him to the local dispensary. Few days after receiving treatment, Lawrence's wound started showing signs of recovery but my condition remained unchanged. I continued with my hot water and olive-oil treatment. Within the last week of his cure, however, my condition started to improve and at a much faster rate than his. Our wounds healed completely by the end of the week.

A few months later my uncle joined the church, presented his idol for destruction by fire, and received baptism. On his annual leave my father visited home. During a worship service he came forward to give testimony about my defiant uncle's conversion. It was only then that I learned the sudden healing of my wound was the reason for my uncle's conversion from paganism to Christian faith. He could not account for my sudden healing other than by supernatural intervention. Regardless of what skeptics may say, I agree with my late uncle that my healing was concrete evidence of the supernatural manifesting itself in the natural world. I shall augment this claim by citing another experience.

As a result of the Biafran War (1967-70), unable to finish my secondary education, I left my hometown for Lagos, where I spent time with two other uncles before I was accepted into a two-year advanced program at the Federal School of Science. As an adult, in a metropolitan city, I found life quite different from my earlier experience. The HGDC at Ajegunle, Lagos, had few members. My social and educational commitments increased, but my zeal for church activities dwindled. Nonetheless, I still attended services occasionally.

At the time I resided in the school's hostel and happened to be assigned one that faced the Atlantic Ocean. Although I grew up in a river environment, I still am fascinated whenever I see an unbounded expanse of water. Even today I still choose a window seat during flights to enable me to gaze in awe as the plane flies over an ocean. During my first week at the hostel, watching the Atlantic Ocean became my hobby. I would return from class to place my books on the window sill and, cradling my jaw in both palms, would reflect on the awesomeness of God.

In retrospect I realize that my conception of God was changing. My understanding was shifting from a personal perspective (Newtonian) to a non-personal perspective (Einsteinian). I pictured how tiny I was against a horizon profiled over an endless expanse of water. I imagined the depth of the ocean, envisioned the numerous creatures there abound, and contemplated the mystery of how God could be concerned about me. I was certain that God exists as an illimitable spirit, but I started to wonder whether he could also be personal. This doubt, of course, was not a denial of God but rather an acknowledgment of my unworthiness. I questioned whether Yahweh, for whom both Moses and Jesus bore testimony, was spiritually with me as I had been taught. My Einsteinian view of God, however, did not interfere with my desire to worship. One thing was clear; in every session of contemplation, I never prayed for an answer.

The HGDC at Ajegunle was several miles from my hostel. Occasionally I returned to my uncles to spend a weekend and attend church services. Several weeks after my new concept about God had established itself, I went to the HGDC church to worship, but no one was there. I waited for a while and then proceeded with the worship service alone. I had received training at home on how to conduct services when ministers were not available.

A few minutes after I began, another member named Money, whom I knew from home, joined me. As we sang and worshipped together, she suddenly came under the Holy Spirit's influence and started prophesying. The experience was not new because I had observed her before when she was filled with the Holy Spirit. This is characteristic of churches that worship God in spirit and in truth. Believers will recall that on the day of Pentecost, the disciples of Jesus were filled with the Holy Spirit and spoke in different languages to a crowd of different nations. With Money, however, what was new on this eventful day was that she called out my name with a special message: "Michael! Michael! The Lord says to you that you doubt in your mind whether I am with you! Rest assured that I am with you always." I was dumbfounded. Only a personal God could tell me exactly what my state of mind was. Given this experience, absolutely nothing will convince me that there is no God! I am confident to say that *my faith anchors on experiential evidence. It is, thus, not blind!*

Writing books was not part of my agenda. I remember another revelation from God while I was at Oyobu. The morning service, which usually commenced at 6:00 a.m., was already in progress when I hurried down the aisle, knelt, said a brief prayer, and then sat down. The chief visionary person, Sunday, was beaming with revelations from the Lord. He suddenly came to me and told the ministers officiating at the altar "These tiny fingers of Michael you see in this natural realm are exceedingly huge in the spiritual realm." I wondered what that meant, but because his description was so vivid it is hard to forget.

Because God exists, I cannot allow either political correctness or my reserved personality to imprison the truth of God's revelation. I am not claiming to be anything but simply a child of God. There are many people with stronger testimonies about God's existence and interventions than mine. People encounter or visualize God in different ways. Sir Isaac Newton conceived of God as a personal being; Albert Einstein perceived God as the "illimitable Spirit." If

Dawkins has not experienced God, it is because he has a closed mind.

In a spirited debate with Christian geneticist Francis Collins, for example, Dawkins said: "The question of whether there exists a supernatural creator, a God, is one of the most important that we have to answer. I think that it is a scientific question. My answer is no."[52] In Dawkins' *The God Delusion*, however, there is insufficient science to justify his answer, prompting scientist and theologian Alister McGrath to remark:

> Curiously, there is surprisingly little scientific analysis in *The God Delusion*. There's a lot of pseudoscientific speculation, linked with wider cultural criticisms of religion, mostly borrowed from older atheist writings . . . . Why were the natural sciences being so abused in an attempt to advance atheist fundamentalism?[53]

McGrath goes on to compare Dawkins' intellectual journey and his own in the following passages:

- Dawkins and I have thus traveled in totally different directions, but for substantially the same reasons. We are both Oxford academics who love the natural sciences. Both of us believe passionately in evidence-based thinking and are critical of those who hold passionate beliefs for inadequate reasons. We would both like to think that we would change our minds about God if the evidence demanded it. Yet, on the basis of our experience and analysis of the same world, we have reached radically different conclusions about God. The comparison between us is instructive, yet it raises some difficult questions for Dawkins.
- Dawkins, who is presently Professor of the Public Understanding of Science at Oxford University, holds that the natural sciences, and especially evolutionary biology, present an intellectual superhighway to

atheism—as they did for him in his youth. In my own case, I started out as an atheist who went on to become a Christian—precisely the reverse of Dawkins' intellectual journey. I had originally intended to spend my life in scientific research but found that my discovery of Christianity led me to study its history and ideas in great depth. I gained my doctorate in molecular biophysics while working in the Oxford laboratories of Professor George Radda, but then gave up active scientific research to study theology.[54]

I close this chapter with the assurance that God is willing to fellowship with anybody who genuinely seeks after the truth. And here is the truth as presented by Robert A. Naumann, Professor of Chemistry and Physics at Princeton University: "I hold that God is the totality of the universe; this includes all scientific principles, all matter and energy, and all life-forms. The existence of the universe requires me to conclude that God exists."[55] The world deserves this TRUTH and not the pseudoscientific speculations presented by the "new atheists."

NOTES

1. *Science and Religion: From Conflict to Conversation* (New York: Paulist Press, 1995), p. 55.
2. *Did Darwin Get It Right?* (Huntington, IN: Our Sunday Visitor, 1998), p. 13.
3. "The Question of Origin Seems Unanswered If We Explore from a Scientific View Alone," *Cosmos, Bios, Theos,* ed. Henry Margenau and Roy Abraham Varghese (La Salle, IL: Open Court, 1992), p. 123.
4. *Science and Religion,* pp. 16, 33.
5. *The God Delusion* (Boston: Houghton Mifflin, 2006), p. 243.
6. Ibid, p. 31.
7. Ibid, pp. 120, 121, 153, 154.
8. Ibid, p. 19.
9. Quoted in "Obituary: Einstein Noted As an Iconoclast in Research, Politics, and Religion," *New York Times,* 19 April 1955, p. 25.
10. *God Is No Delusion* (San Francisco: Ignatius Press, 2007), p. 28.
11. Ibid, pp. 31-32.
12. Ibid, p. 45.
13. *The Blind Watchmaker* (London: Penguin, 2006), p. 14.
14. "The God Delusion? (Part 6)." intentBlog. 17 November 2006. www.deepakchopra.com.
15. *Creation: Facts of Life* (Green Forest, AR: Master Books, 2006), p. 122.
16. "Cargo[-]Cult Science," *Surely You're Joking, Mr. Feynman!: Adventures of a Curious Character* (New York: W. W. Norton, 1985), p. 343.
17. Ibid, p. 341.
18. "The Information Challenge." http://www.noanswersingenesis.org.au/ dawkinschallenge.htm. Retrieved 8 August 2008.
19. *Information Theory, Evolution, and the Origin of Life* (Cambridge: Cambridge University Press, 2005), p. 188.
20. *The God Delusion,* p. 79.
21. Ibid, p. 117.

22. *Creation: Facts of Life*, pp. 122-23.

23. *The Blind Watchmaker*, p. 141.

24. Ibid.

25. *The God Delusion*, p. 111.

26. *The Dawkins Delusion?: Atheist Fundamentalism and the Denial of the Divine* (Downers Grove, IL: InterVarsity Press, 2007), p. 12.

27. *The God Delusion*, p. 99.

28. Ibid, p. 102.

29. *The Dawkins Delusion?*, p. 44.

30. "A Scientist of Big Ideas Thanks His Lucky Stars." *Globe and Mail,* 7 April 2011, A3.

31. *There Is a God: How the World's Most Notorious Atheist Changed His Mind* (New York: HarperCollins, 2007), p. xxiv.

32. *The Dawkins Delusion?*, p. 18.

33. *The God Delusion*, p. 338.

34. Ibid, p. 68.

35. *There Is a God*, dust jacket.

36. Ibid, pp. 123-24.

37. Ibid, p. 129.

38. Ibid, pp. 88-89.

39. Ibid, p. 93.

40. Ibid, p. 89.

41. Ibid, pp. 90-91.

42. "Science Asks What and How, While Religion Asks Why," *Cosmos, Bios, Theos*, p. 84.

43. *There Is a God*, pp. 78-80.

44. Ibid, p. 111.

45. Ibid, p. 124.

46. *What Darwin Got Wrong* (New York: Farrar, Straus and Giroux, 2010), p. xx.

47. *There Is a God*, p. 131.

48. Ibid, p. 132.

49. "Why the Existence of God Is Not Required to Explain the Existence of the Universe," *Cosmos, Bios, Theos*, p. 238.

50. *Ideas and Opinions*, trans. Sonja Bargmann (New York: Dell, 1973), p. 255.

51. *The God Delusion*, pp. 87-88.
52. Quoted in "God vs. Science," *Time*, 13 November 2006, p. 35.
53. *The Dawkins Delusion?*, pp. 11-12.
54. Ibid, pp. 9-10.
55. "Religion and Science Both Proceed from Acts of Faith," *Cosmos, Bios, Theos*, p. 72.

CHAPTER 10

THE MEDIA AND THE CREATIONISM-EVOLUTIONISM CONTROVERSY

I have found it practically impossible . . . to get newspapers to acknowledge that there are scientific problems with Darwinism that are quite independent of what anybody thinks about the Bible. A reporter may seem to get the point during an interview, but after the story goes through the editors it almost always comes back with the same formula: creationists are trying to substitute Genesis for the science textbook. Scientific journals follow the same practice.[1]

—Philip E. Johnson

The director of education at Britain's Royal Society has been forced to resign after a massive outcry in the wake of *widespread misreporting of comments he made about creationism* in the classroom. Michael Reiss, a professor at London's Institute of Education, . . . made the remarks at the British Association for the Advancement of Science's annual Festival of Science on 11 September in Liverpool.

Three Nobel-prizewinning society fellows wrote to the society's president, Martin Rees, saying that they were "greatly concerned" at *media reports of Reiss's talk* . . . . After the letter of complaint and with the reported statements continuing to receive *press-coverage, including hostile opinion pieces*, the society announced Reiss's departure on 16 September.[2] (Emphasis mine.)

—*Nature*, September 2008

DESERVED RESPECT FROM THE MEDIA

The media's prejudice against creationism is transparent. The media's latest misinterpretation concerns the views of Professor Michael Reiss, Director of Education for the Royal Society, which culminated in his resignation. In the editorial section under the title "Creation and Classrooms," *Nature* reported as follows:

The headlines were damning. "Leading scientist urges teaching of creationism in schools," proclaimed Britain's *The Times* newspaper on 12 September, echoing the headlines appearing that day in numerous other British media. The stories asserted that Michael Reiss, a biologist and educational researcher, an ordained Anglican minister and (at the time) the education director of the Royal Society, had explicitly advocated that state-school biology classes teach creationism.

The reports were wrong. Speaking at the British Association for the Advancement of Science's annual Festival of Science on 11 September, Reiss had articulated—as he had many times before—a view consistent with the Royal Society's official position: when students from a creationist background raise the issue in class, the teacher should explain why creationism is not science and why evolution is. Nevertheless, on 16 September the society announced Reiss's departure, arguing that the media's misinterpretation had "led to damage to the society's reputation."[3]

The Royal Society is the United Kingdom's National Academy of Science. Reiss was pressured to resign because of his comments that science teachers should treat creationists' beliefs "not as a misconception but as a worldview."[4] Reiss was simply affirming a statement of fact! Creation by God is the official worldview of the United Kingdom and the Commonwealth nations since in parliamentary assemblies they pledge, "God Save the King/Queen." A survey timed to coincide with the bicentenary of Charles Darwin's birth showed that half of Britons reject evolution.[5] Reiss thus was only being realistic. Here are some of his remarks:

> My experience after having tried to teach biology for 20 years is if one simply gives the impression that such children are wrong, then they are not likely to learn much about the science. I realized that simply banging on about evolution and natural selection didn't lead some pupils to change their minds at all. Just because something lacks scientific support doesn't seem to me a sufficient reason to omit it from the science lesson.[6]

As long as educators shy away from making a distinction between bacteria-to-bacteria evolution that can be demonstrated (science) and bacteria-to-human evolution that can never be demonstrated (belief), they do a great injustice to students and the public. This clarification suggests that the main reason for promoting only evolution by natural selection is to persuade the public to embrace only evolutionism. Reiss's above comments confirm that the goal in the science classroom is to cause students change their minds from a creationist to an evolutionist worldview. This is morally and academically objectionable.

Richard Dawkins has made it abundantly clear in his public speeches and in his book titled *The God Delusion* that the teaching of evolutionism leads to atheism. However, as Hubert Yockey points out, "Theism and atheism both are irrelevant to science because they address problems of faith and belief."[7] Dawkins is a Fellow of the Royal Society and Charles Simonyi Professor of the Public Understanding of Science at Oxford University. Presumably because of its strong atheistic membership, the Royal Society did not react to Dawkins' declaration that evolutionism influences students to become atheists. Moreover, Oxford's administration was unconcerned that its endowed professor responsible for promoting the public understanding of science had divagated into championing atheism.

The Times was not critical of Dawkins' comments but was swift to denounce Reiss's defense of creationism in the science classroom: "To consider creationism and its stepchild intelligent design as if they were science is to inflict an injustice on schoolchildren."[8] The institutionalized hypocrisy I see here lies in teaching schoolchildren to sing about God as Creator in the national anthem and then insisting that schoolchildren accept evolutionism as a scientific fact without proof. Furthermore, the decision to relieve the Royal Society's Director of Education of his post for defending the teaching of creationism as a worldview is additional evidence of atheistic evolutionists' intolerance. Its reaction is no more civil than the ones orchestrated by religious fanatics or fundamentalists.

To subject students to a single worldview in the science classroom is to deny them the right to freedom of choice. The bacteria-to-human version of evolution has not been established, and, as is also true of its rival special creation, it will never be established in any child's lifetime. Science is based on empirical facts, not on improbable speculation. Anything that science cannot demonstrate to the satisfaction of *every accomplished scientist* is not a scientific fact. On both sides of the creationism-evolutionism controversy, however, we have Nobel Prize laureates, and theories on the origin of species divide the scientific community. Yet we live in a world where advocating atheistic sentiments in science is seen as a measure of brilliance and academic prowess. It is, of course, wrong to link atheism to academic achievement. Sir Isaac Newton, who was President of the Royal Society from 1703 to 1727, pronounced his scientific discoveries as evidence of God's existence.[9] Newton shunned atheism as nonsensical. In the present row over creationism, his views are exonerated. The remarks of Lord Winston, Professor of Science and Society at Imperial College,

London, are noteworthy: "I fear that the Royal Society may have only diminished itself. This individual [Michael Reiss] was arguing that we should engage with and address public misconceptions about science—something that the Royal Society should applaud."[10] The Royal Society should allow teachers to present evolutionism alongside creationism and let students figure out for themselves, with the guidance of teachers as facilitators, which is scientifically more plausible. John MacArthur points out in *The Battle for the Beginning:Creation, Evolution, and the Bible*:

> The conviction that nothing happens supernaturally is a tenet of faith, not a fact that can be verified by any scientific means. Indeed, an *a priori* rejection of everything supernatural involves a giant, irrational leap of faith. So the presuppositions of atheistic naturalism are actually no more "scientific" than the beliefs of biblical Christianity.[11]

If the science classroom is deemed not an inappropriate forum for comparative studies, then transfer all worldviews on origin to the world-religion classroom. *It is wrong to use science in any form to promote an atheistic worldview.*

As things now stand, Dawkins' motive, endorsed by the Royal Society and most media, is to convince schoolchildren that the God they revere in reciting the national anthem is a myth. Interestingly, however, the popular rating for Dawkins' *The God Delusion* was 59 on a scale of 0 (worst) to 100 (best). Some 22 critical reviews of the book yielded the following results: outstanding (3), favorable (6), mixed (9), and unfavorable (4).[12] This disparity of response to Dawkins' views suggests a revealing split between "official" endorsement of his ideological stance and the reading public's reception of Dawkins' case for monolithic education.

Since most of the leading scientists in our highly secularized world are evolutionists, the media regard their views as final and dismiss the views of accomplished creationists as religious propaganda. For instance, in 2006 Polish scientist Maciej Giertych complained about media bias in a letter to *Nature*:

SIR—In your news story "Polish Scientists Fight Creationism" (*Nature* 443, 890-91, 2006), you incorrectly state that I have called for the "inclusion of creationism in Polish biology curricula." As well as being a member of the European Parliament, I am a scientist—a population geneticist with a degree from Oxford University and a Ph.D. from the University of Toronto—and I am critical of the theory of evolution as a scientist, with no religious connotation. It is the media that prefer to consider my comments as religiously inspired, rather than to report my stated position accurately. I believe that, as a result of media bias, there seems to be total ignorance of new scientific evidence against the theory of evolution.[13]

This does not come as a total surprise, since members of the media have their personal philosophical preferences to promote.

The media not only misinterpret views that are anti-evolutionism, but they also promote evolutionism in subtle ways. For instance, chemist E. C. Ashby expressed concern about the 18 June 2003 television program titled *Walking with Cavemen* in which apes like Lucy, whom evolutionists wrongly identified as the missing link, are portrayed as thinking humans. "What a shame," he writes, "that an attempt is still being made by the media to deceive the public with regard to the acceptance of evolution."[14]

Conventionally the media is expected to be neutral. Facts should be reported simply as facts; interpretation should be left to the public. Wallace Johnson describes the media's bias as follows: "The ranks of anti-evolution scientists are growing; but the mass media ignores [*sic*] them, or discredits [*sic*] them by disparagement."[15] It seems, on the one hand, that the media applaud atheistic fundamentalism; on the other hand, they disparage theistic fundamentalism. The media's anti-God reaction is consistent with Deepak Chopra's observation that the media often follow Dawkins' lead:

Dawkins has written extensively on evolution . . . and speaks out loudly against any possibility that God is real . . . . Eventually science will uncover all mysteries. Those that it can't explain don't exist . . . . Dawkins makes it an us-versus-them issue. Either you are for science (that is, reason, progress, modernism, optimism about the future) or you are for religion (that is, unreason, reactionary resistance to progress, clinging to mysteries that only God can solve) . . . . Sadly, the media often follow his lead, erasing the truth, which is that many scientists are religious and many of the greatest scientists (including Newton and Einstein) probed deep into the existence of God.[16]

I wholeheartedly affirm Chopra's comments based upon my recent experience in Toronto.

Toward the end of June 2007, Dawkins was in Toronto to deliver a lecture on the irrationality of religious belief at the Idea City conference. At the time his book *The God Delusion* was among the best-sellers worldwide. Three local newspapers—*Toronto Star*, *Globe and Mail*, and *National Post*—covered his presentation, which was entertaining and informative concerning his atheistic view of life. Two articles in the *Toronto Star*, one by Stuart Laidlaw (faith and ethics reporter) and the other by Tom Harpur (special to the *Star*), appeared on the same page.[17]

Laidlaw's piece, "No Rotten Tomatoes for Anti-God Author," featured supporting statements from Isabel Mattson, Head of the Islamic Society of North America, who surprisingly endorsed Dawkins' views by asserting that "Much of what we say is God is a delusion." The article also cited Jay Bakker, son of former TV evangelists Jim and Tammy Faye Bakker, who told Dawkins that a few atheists who attend his Revolution Church have asked him to read *The God Delusion*.

Harpur's "Behind the Atheist Upsurge" advertised ten best-selling anti-God books, with either the author's name or the book title printed in the corner. The titles included *Without God, How Religion Poisons Everything, The End of Faith, Letter to a Christian Nation, God Is Not Great, The God Delusion*, and *The Atheist's Bible*. The authors who were represented included Sam Harris,

Christopher Hitchens, and Richard Dawkins. Harpur's article also listed quotations by such alleged non-believers as Abraham Lincoln, Albert Einstein, Aldous Huxley, Isaac Asimov, George Carlin, and Gloria Steinem.

Einstein, however, was never an atheist! There is a remarkable difference between believing in God, not accepting certain concepts about God (Einstein belongs to this class), and not believing in God in any form (Dawkins belongs here). Confirming Einstein's discontent at being classified as an atheist, biographer Walter Isaacson asserts:

> [T]hroughout his life, Einstein was consistent in rejecting the charge that he was an atheist. "There are people who say there is no God," he told a friend. "But what makes me really angry is that they quote me for support of such views." . . . Unlike Sigmund Freud or Bertrand Russell or George Bernard Shaw, Einstein never felt the urge to denigrate those who believed in God. Instead, he tended to denigrate atheists. "What separates me from most so-called atheists is a feeling of utter humility toward the unattainable secrets of the harmony of the cosmos," he explained.[18]

On religious matters, Einstein was thus the exact opposite of Dawkins. Because Einstein insisted that he was not an atheist, he cannot be ranked among such die-hard types as Hitchens and Dawkins. Too often atheists like Dawkins quote only parts of Einstein's pronouncements that appear to favour their philosophical belief, thereby misleading the public. This tendency confirms Einstein's statement that fanatical atheists are "creatures who—in their grudge against traditional religion as the 'opium of the masses'—cannot hear the Music of the Spheres."[19] If atheism is the belief that God does not exist, Einstein is not an atheist.

Jon Allemang of the *Globe and Mail* reported on Dawkins' two-day mission to Toronto under the title "The Infinite Wisdom of Richard Dawkins."[20] Joseph Brean of the *National Post* also reported on Dawkins' lecture under the title "Celebrity Deity-Slayer

Has a Bad Day," with the subtitle "Evolutionary Biologist Professes Belief in Aliens."[21] It is evident that the major newspapers in Toronto, except for the *Toronto Sun*, gave thorough coverage to Dawkins' lecture, but it appears their general objective was to market atheism to the Canadian public. For the *Globe and Mail* Dawkins is a man of "infinite wisdom," and it is no accident that the newspaper was one of six worldwide to print a favorable review of *The God Delusion*.[22] But would Dawkins' mission, boosted by media support, change the minds of 42% of the Canadian population, which, according to an Angus Reid poll released prior to his visit, believes that dinosaurs roamed the planet alongside humans in recent history?[23]

The Reid poll was conducted on 12-13 June 2007 and involved online interviews with 1,088 Canadian adults. Dawkins then visited Toronto during the last week of June. On the morning of 4 July, while on my way to work, I picked up two local newspaper briefs, *Metro* and *Toronto 24 Hours*, which are good summaries of the Canadian press. I was fascinated by the Decima Research survey results in both papers, which showed that belief in creation was twice that of belief in evolution.[24] *Toronto 24 Hours* gave the results of this nationwide survey in great detail, featuring on the front page a story titled "Creation Tops Poll" and subtitled "Most See God Having Essential Role in Creating Mankind." The article continued on page 10 under the title "Creation Tops Evolution in New Poll." An excerpt reads:

- Canadians may not be as religious as Americans, but a new poll suggests they are not prepared to rule out God's essential role in creation.
- The Canadian Press-Decima Research survey suggests that 60 per cent of Canadians believe God had either a direct or indirect role in creating mankind, shattering the myth that Canadians had long ago put their faith strictly behind the scientific explanation for creation.
- The poll suggests Canadians divide in essentially three groups on the issue of creation: 34 percent of those

polled said humans developed over millions of years under a process guided by God; 26 percent said God created humans alone within the last 10,000 years or so; and 29 percent said they believe evolution occurred with no help from God.

- "These results reflect an essential Canadian tendency," said pollster Bruce Anderson. "We are pretty secular, but pretty hesitant to embrace atheism."

Given the fact that the *Toronto Star*, *Globe and Mail*, and *National Post* covered Dawkins' anti-religion crusade, I expected that at least some, if not all, would publish the Decima Research survey results. I needed this information for future reference, so I went searching for it in the major newspapers. I was dumbfounded when I realized that not even one of them had published this poll's results on 4 July, the same day the local briefs carried the news. I hoped that the weekend editions might carry it, as some had done with Dawkins' anti-God campaign. Searching from the 4th through the 8th of July, however, I found nothing. Then it dawned on me that it probably was deemed unworthy of attention, though Dawkins' lecture had received elaborate coverage. It was only then that I started to reflect on creationists' allegations of media bias. I recalled the experience of Phillip E. Johnson, a law professor and author of *Darwin on Trial*, with media bias (see epigraph at the beginning of this chapter).

The news that most Canadians believe God had a direct or indirect role in creation was inconsequential to the editors of major Canadian newspapers. Presumably they felt that publishing such poll results would be seen as supporting creationists' point of view and making Canada appear behind the times in scientific awareness. What other reason could there be for the consistency among the major newspapers in covering Dawkins' address and avoiding the news that creationism (theism) prevails over evolutionism (atheism)?

In order to maintain the status quo, about a month following Dawkins' visit another article was published on the front page of the *National Post* titled "Ditching God: Emboldened Atheists

Are Finding Purpose in Coming Out of the Closet." It continued on page A8 under the subtitle "Stigmatized No More." The piece offers these statistics:

- A City University of New York survey found the number of non-religious adults grew from 8% to 14.3% between 1990 and 2001, to more than 29 million Americans. The current issue of *The Atlantic* magazine cites a study that showed 14% of Americans "were distancing themselves from organized religion as a symbolic gesture against the religious right." A 2006 Pew study found that 20% of today's 18- to 25-year-olds years have no religious affiliation or are atheist or agnostic, up from 11% in the late 1980s.
- In Canada, the number of people who categorize themselves as atheists, agnostics, humanists or no-religion rose to 16.2% in the 2001 census, up from 12.3% in 1991, and 7.4% a decade earlier.[25]

The report included a bar graph showing "trends in religious membership" covering 1981, 1991, and 2001. The graph shows declines in both Catholic and Protestant membership but an increase for agnostics and atheists. Highlighted in bold italics below the graph is the caption, "*It is very comforting to know your neighbour . . . is an atheist, too.*" This article is clearly a media promotion of atheism.

The Canadian media's support for evolutionism is obvious, as it surfaced again in the 2007 provincial election. The media wasted no time in responding when, during the first week of September 2007, Ontario Progressive Conservative leader John Tory contended that creationism should be taught alongside evolutionism in publicly funded schools, for doing so would ensure that students received a well-rounded education. All the major newspapers carried this story, some on the front page, because the proposal was seen as a violation of the curriculum. The furor subsided only when Tory retracted his comment and agreed to an evolution-only science curriculum. The *Toronto Star* reported: "He was later forced to issue clarification that his proposal would allow creationism to be discussed only as part of religious programming, as is now the practice in Ontario's publicly funded Catholic schools, and not in science classes."[26]

Since when has the Darwinian myth become the entrenched gospel of nations such that the media police its acceptance as a scientific fact? The media, at the very least, should reexamine the creationism-evolutionism controversy in relation to public education. In a recent *TIME* magazine article, "The Case for Teaching the Bible," David Van Biema writes: "Should the Holy Book be on the public school menu? Yes. It's the bedrock of Western culture. And it's constitutional—as long as we teach but don't preach it."[27]

For those who take the Canadian National Anthem seriously, the case for banning creationism and teaching only evolutionism in science classrooms raises some questions. On the one hand, the Constitution requires the pledge, "God keep our land glorious and free"; on the other hand, the educational curriculum demands that God must be banned from the science classroom. How can I in good conscience teach evolutionism when I know that it, like creationism, does not meet the requirements of science? The situation is tantamount to indoctrination masquerading as education.

The battle being waged, using science as the bandwagon, involves nothing less than the elimination of God from our understanding of the world and how it came into existence. Why insist on Darwinian evolution by natural selection, a theory that cannot be proven? It is reprehensible that those who reject atheistic materialism are marginalized, stereotyped, and subjected to media bashing. This must stop. My views do not imply that all persons involved in the media are guilty of aiding and abetting Dawkins' vision of a world devoid of religion. Some undoubtedly do not subscribe to Dawkins' perverse brand of fundamentalism, and I respect their views.

In sum, two forces are intent on advancing the paradigm of Darwinian evolution. First, in order to indoctrinate the public, leading evolutionists present their philosophical opinions as scientific facts while downplaying the social impact of their claims.

Second, the media frustrate the efforts of creationists to point out problems inherent in Darwinian evolution by natural selection. The media in this regard oppose academic freedom. Under these circumstances *The Darwinian Delusion* is appropriate for people who honestly seek the truth.

NOTES

1. *Defeating Darwinism by Opening Minds* (Downers Grove, IL: InterVarsity Press, 1997), p. 34.
2. "Creationism Row Forces Out UK Educator," *Nature*, September 2008, p. 441.
3. "Creation and Classrooms: Better to Confront Superstition Science Than to Disregard the Superstition," *Nature*, September 2008, p. 431.
4. Quoted in Lewis Smith and Mark Henderson, "Royal Society's Michael Reiss Resigns Over Creationism Row," *The Times*, 17 September 2008. http://www.timesonline.co.uk/tol/news/uk/science/article4768820.ece.
5. Riazat Butt, "Half of Britons Do Not Believe in Evolution, Survey Finds." http://www.guardian.co.uk/science/2009/feb/01/evolution-darwin-survey-creationism.
6. Quoted in Smith and Henderson.
7. "Message from Professor Hubert Yockey," *Truth Journal.* http://www.leaderu.com/truth/1truth18c.html.
8. BBC News, "Call for Creationism in Science," 13 September 2008. http://news.bbc.co.uk/2/hi/uk_news/education/7612152.stm.
9. See Wikipedia, "Isaac Newton's Religious Views." http://en.wikipedia.org/wiki/ Isaac_Newton's_religious_views; also Ann Lamont, *21 Great Scientists Who Believed the Bible* (Acacia Ridge, CA: Creation Science Foundation, 1995), pp. 37, 47.
10. Quoted in Smith and Henderson.
11. *The Battle for the Beginning: Creation, Evolution, and the Bible* (Nashville: W Publishing Group, 2001), pp. 50-51.
12. Metacriti.com Books, *The God Delusion* by Richard Dawkins, Overall Metascore, http://www.metacritic.com/print/books/authors/dawkinsrichard/goddelusion.
13. "Creationism, Evolution: Nothing Has Been Proved," *Nature*, 16 November 2006, p. 265.
14. *Understanding the Creation-Evolution Controversy* (Ozark, AL: ACW Press, 2005), p. 51.

15. *The Death of Evolution* (Rockford, Illinois: Tan Books, 1986), p. 3.
16. "The God Delusion? Part 1," intentBlog, 15 November 2006. www.deepakchopra.com.
17. *Toronto Star*, 23 June 2007, p. ID6.
18. "Einstein and Faith," *Time*, 16 April 16 2007, p. 35.
19. Ibid, pp. 35-36.
20. "The Infinite Wisdom of Richard Dawkins," *Globe and Mail*, 23 June 2007, Focus F3.
21. "Celebrity Deity-Slayer Has a Bad Day," *National Post*, 21 June 2007, p. A10.
22. Metacriti.com Books, *The God Delusion* by Richard Dawkins, Overall Metascore, http://www.metacritic.com/print/books/authors/dawkinsrichard/goddelusion.
23. Angus Reid Global Monitor, "Most Canadians Pick Evolution Over Creationism," http://www.angus-reid.com/polls/view/16178.
24. Bill McDonald, "Most Believe God Had Role in Creation," *Metro*, 4 July 2007, p. 6; and Sun Media Corporation, "Creation Tops Poll," *Toronto 24 Hours*, 4 July 2007, pp. 1, 10.
25. Charles Lewis, "Ditching God: Emboldened Atheists Are Finding Purpose in Coming Out of the Closet," *National Post*, 21 July 2007, pp. A1, A8.
26. Richard Brennan, "Creationism Could Be Taught in Funded Schools, Tory Says," *Toronto Star*, 6 September 2007, p. A15.
27. "The Case for Teaching the Bible," *TIME*, Canadian Edition, 2 April 2007, p. 28.

CONCLUSION

If uniformitarianism is denied, all of science becomes impossible (*The Problems of Evolution,* Oxford University Press, 1985, page 7).

Mark Ridley

No one has proved experimentally the idea that large variations can emerge from simpler life forms in an unbroken ascendancy to man. A large body of scientific evidence in biology, geology and chemistry, as well as the fundamentals of information theory, strongly suggest that evolution is not the best scientific model to fit the data that we observe.

Concerned Scientists and Educationists
[See Appendix A for 27 signatories]

There is no creation-evolution controversy; the controversy is over creationism and evolutionism, which are dogmas that fall short of the scientific requirements of testing, retesting and experimentation. Creationism portrays the normal scenario where the designer claims the designed. Evolutionism advocates the abnormal scenario where the designed claims the designer. Science favours creationism; pseudoscience promotes evolutionism.

Michael Ebifegha

Most biologists now accept that the tree is not a fact of nature—it is something we impose on nature in an attempt to make the task of understanding it more tractable. Darwin was wrong.

NewScientist, Editorial and Cover Story,
January 24-30, 2009

Science and religion share a common goal—the commitment to reveal and explain, with integrity, the truth about nature. This book addresses two objectives: first, to clear the materialistic roadblocks that diehard evolutionists have erected on the path to truth; and, second, to argue that God is the totality of truth.

I explained in this discourse that creation parallels evolution as processes of science, while creationism parallels evolutionism as philosophical beliefs/models of origins and diversity of life. The phrase "creation/evolution controversy" is thus misleading because the actual debate is between creationism and evolutionism. I then proposed the use of "intraevolution" and "extraevolution" instead of "microevolution" and "macroevolution" in an effort to isolate evolution (science) from evolutionism (belief/religion). The terms "microevolution" and "macroevolution", currently used to described the bacteria-to-bacteria and bacteria-to-human evolution which are autonomous events, give the false impression that one leads to the other in due course.

On the sensitive question of "how old the Earth is," science and religion constitute different but equally valid ways of viewing the world. Religion is concerned with the structural age, when the Earth as a planet was structured to accommodate life, while science addresses the Earth's matrix age, which is the age of its constituents. Religion addresses the architectural age, but science estimates the geological age. There is no essential conflict between these views.

However, on the question of how life originated and developed, science and religion cannot be viewed as radically different but equally valid ways of understanding the world, since there can be only one truth. According to Einstein, "science without religion is lame[;] religion without science is blind." Neither can exclusively provide a complete understanding of events in an unknown past. Unfortunately, in our modern world, materialism overrules religious paradigms. *Scientists*, therefore, as opposed to science and religion, disagree on philosophical grounds in categorizing themselves as creationists and evolutionists.

On the question of origins, the Judaeo-Christian God is the only supernatural Being who publicly claimed credit for creating the earth and heavens and equipping them with various organisms. Empirical science, through the natural law of biogenesis, corroborates God's claim in stipulating that life cannot emerge spontaneously from non-life but must come from preexisting life. Natural laws are immutably timeless throughout Earth's history; we thus cannot ignore any of them for philosophical reasons.

Science relies on the premise of uniformitarianism, which implies that natural processes are immutable; the processes that were at work in the past are the same as are operative today. According to evolutionist Mark Ridley, "If uniformitarianism is denied, all of science becomes impossible." Therefore, any natural law or hypothesis that is nonexistent today is a betrayal of science in that it constitutes only a *naturalism of the gaps*. Creationists, accordingly, stick to the traditional mandate of science in interpreting evidence. Under that mandate scientists follow the evidence wherever it may lead. For instance, in the case of DNA the fact is that digital programs do not arise by chance and that their development is not through mindless processes. Since the genetic program is far superior in every respect to any of our existing programs, it can only be attributed, analogically, to superior intelligence. Creationists, therefore, following the scientific evidence, believe that there must be a Creator, but they differ concerning the Creator's nature and the manner of creation. Their combining of science and religion is referred to in this discourse as creationism. On the other hand, evolutionists endorse the creed of materialism, according to which everything must be explained only in terms of matter. Evolutionists, accordingly, argue that the genetic program arose spontaneously from matter under the right conditions. On the point of life's origin, evolutionists discount the scientific law of biogenesis because it points to a Creator.

Instead, evolutionists establish the hypothesis of abiogenesis, which postulates that life can originate from non-living matter, a belief that contradicts our experiential knowledge and the law of biogenesis. However, while we can modify or discard scientific theories if they do not fit the empirical evidence, scientific laws are fixed. Rather than abandon the Darwinian theory of evolution, leading modern scientists, in order to eliminate God from the equation, ignore or circumvent the natural laws that contravene the evolutionary worldview. Since science cannot be built on myths, evolutionists have established an "Origin-of-Life Prize" through the Abiogenesis Foundation to award one million dollars to anyone who, based on non-supernaturalistic processes, can theorize a plausible mechanism for the spontaneous emergence of genetic instructions in nature. This example of deliberately conflating science with naturalistic beliefs/preferences in order to ratify an evolutionary worldview is what this book refers to as evolutionism.

Only in the spirit of "Darwinian delusion" would leading scientists publicly declare evolutionism a scientific fact while behind the scenes devoting a million dollars to soliciting evidence to cover the lie. Morally speaking, this is simple fraud. By declaring evolutionism a scientific fact, evolutionists take undue advantage of creationists and thrust the public into doubting the existence of God. Interestingly, the Abiogenesis Foundation has decided not to advertise its "Origin-of-Life Prize" in lay media, preferring to keep it a secret within the scientific community until after the Prize is won. The project has now existed for over ten years without a winner, and was tentatively on hold because its authenticity faces a challenge.

Although it is true that several creationist worldviews exist, the biblical account overshadows all others because it is the cornerstone of world history and, unlike many competing versions, posits a veridical claimant. Evolutionism, born of the denial of a creationist worldview, falls short of the scientific requirements of testability and repeatability. The formal endorsement in 1864 of the biblical creationist worldview by the Philosophical Society of Great Britain, a body of scientists that included 86 Fellows of the Royal Society, constitutes a major event in the history of science. Between that time and now there has been no scientific breakthrough in evolutionary biology. The only major changes within the modern scientific community are the steady growth in the atheistic/agnostic population of scientists in the National Academy of Sciences and Royal Society, and the subsequent rise in the number of pseudoscientific speculations and assertions.

Physicist and information theorist Hubert Yockey's impact on the creationism/evolutionism controversy is noteworthy. Yockey presents data that support creationism, but he offers interpretations that justify evolutionism presumably because he rejects creationism and intelligent design. For instance, comparing living matter and machines, Yockey argues that the role of an intelligent designer in the functioning of machines is accomplished in living matter by the genome; he then concludes that there is no need for an intelligent designer. Yockey's argument founders when the focus is on the "origin of the genome." The human body is itself a machine with the genetic code operating as the machine code, so that the whole system is the product of an intelligent agency. Yockey allows his philosophical beliefs to interfere with objective truth, as I will illustrate with a few other examples.

Yockey contends that messages in the DNA sequences are similar to programs that run modern computers. Now, in terms of design considerations, the origin of computing machines is unequivocally tied to human intelligence. Similarly, the origin of the genome can also be tied to non-material intelligence. This conclusion is anathema to scientists but logically valid on the grounds of scientific analogy. Yockey shows exemplary integrity in his acquisition of data but succumbs to philosophical preference in his interpretation of the data.

Yockey is also the first scientist to define the distinction between living and non-living matter, and he points out that the origin of life is unknowable as a scientific problem. By simple deductive reasoning, if the origin of life is unknowable, the origin of species is undecidable. However, rather than confronting Darwinian theory with this fact, Yockey, argues that the fact of life origination should simply be taken as an axiom, or a truth that cannot be proven.

If this is the consensus, it follows that the origin of species can only be treated similarly as an axiom, and under such proviso the Darwinian paradigm is reduced to bacteria-to-bacteria evolution only. Accordingly, Yockey is right in his conclusion that there is no need for records of intermediate structures but flounders in his assertion that Darwin's theory of evolution is as well-established as any theory in science. Yockey's stance clearly confirms physicist H. S. Lipson's insight that evolutionism is a scientific religion and that many scientists are prepared to "bend" their observations to fit it. Science has nothing to do with preference, but it has everything to do with the objective truth.

Religion and science agree on the diversity of life forms but express their views in different ways. According to the Scriptures, God fashioned the earth, sea, and heavens, using clay for the various terrestrial components. Although the different life forms were composed of the same material, they were created separately and endowed with dissimilar material and immaterial qualities that science has been unable to address until now. The organisms were made to diversify through reproduction, each replicating its own kind. According to God's calendar of Creation, human beings were the last of all organisms to come into existence and were specifically made in the image of God but endowed with free will. They were given dominion over other organisms and mandated to subdue the earth scientifically and technologically. Human beings, accordingly, had a special relationship with God, which continues to this day. The formal breakdown of human/ divine fellowship came when human beings, acting on their free will, compromised their trust in God. At the appropriate time in the history of the world, God intervened to issue the Ten Commandments for holy living to the ancient Israelites. God's claim to have created the universe was presented concomitantly with these moral commandments. In the wake of a damaged relationship, corruption crept in, and since then the universe has been subjected to constant deterioration in these forms: (1) the inanimate realm, such as the earth's structure, became subject to earthquakes, tsunamis, and erosion; and (2) the animate realm, exposed to the contamination of genetic programs, became vulnerable to copying errors in the blueprint of life. Changing environments and other factors such as natural selection produced significant variations within various species but not transmutations between them. The boundaries between life forms remain fixed, as is evident in the fossil record and the living world. For over 150 years of intense research studies, scientists are only able to demonstrate the limited changes or evolution within a given kind of organisms.

Science corroborates religion regarding the homologous similarities among organisms, irrespective of classification. It reveals that genetic programs are similar in complexity, meaning that organisms could not have progressed from simple to complex in the course of Earth's history. Such evidence logically supports the scenario in which the basic organisms were created over short time intervals as opposed to millions of years. Science also empirically justifies the bacteria-to-bacteria form of evolution, consistent with the religious assertion that organisms and plants can reproduce only after their own kind. Variations thus are confined to respective boundaries. Science, however, is unable to demonstrate any bacteria-to-human form of evolution.

The disagreement on the designing agency within the scientific community is a result of different interpretations of the same evidence by creationists and evolutionists. For creationists the similarities among organisms affirm that an Intelligent Designer who used the same material for diverse purposes created various life forms. Evolutionists, on the other hand, insist on an exclusively materialist explanation, contending that the apparent similarities are evidence of a common ancestry originating by chance as opposed to an Intelligent Designer. Although this is a valid philosophical argument regarding similarities, it cannot account for great dissimilarities among the various organisms. While a Creator can impose any number of dissimilarities among organisms, inanimate matter, aided by mindless natural processes, cannot develop the remarkable array of dissimilarities among life forms.

Members of the scientific community disagree in their interpretation of the facts of science and religion, although they concur that evolution operates as a natural process. The controversy concerns the extent to which evolution influences biological systems. No scientist questions bacteria-to-bacteria evolution, which is the focus of ongoing applications in medicine and agriculture, but bacteria-to-human evolution is widely disputed. Evolutionists and creationists alike express their commitment to scientific methodology; however, neither camp can demonstrate the truth of their beliefs. Evolutionists suggest that we need to wait million of years for major evolutionary change to materialize, even though the "wait and see" argument is antithetical to science. Scientific evidence from palaeontology has consistently revealed the fallacy of this evolutionist postulate. In reality, only a collection of fossils of various organisms from the same geological location can provide concrete natural documentary record of either creation or evolution; isolated fossil fragments such as *Archaeopteryx*, *Toumai* and *Tiktaalik,* may be described as compelling but are certainly not proofs of any worldview. As a witness against evolutionists, however, are two deposits, one at the Burgess Shale in the Canadian Rockies and the other at Chengjiang in China, where fossils of many anatomically and genetically different organisms appear fully formed with no intermediate stages. Here, we have in our hands concrete historical and scientific documentary evidence for which the creationist worldview scores 2 points and the evolutionist worldview has no points! Small wonder evolutionist Mark Ridley, asserts, "No real evolutionist, whether gradualist or punctuationist, uses the fossil record as evidence in favour of the theory of evolution as opposed to special creation." Nature, as though further to embarrass skeptics, preserves the platypus, a monotreme with avian, reptilian, and mammalian traits. In over 100 million years of its existence according to the evolutionary time-scale, the ancient platypus fossil and the modern adult platypus display only intraevolutionary (microevolutionary) changes consistent with those observed in animal breeding. That is, over the span of 100 million years we observe only platypus-to-platypus modifications. The same can be said of the turtle that shows no evidence of transitional stages for over 200 million years on the evolutionary time-scale. Evolution by natural selection is consistent with that observed in artificial selection and, furthermore, is congruent with the scriptural account of limits imposed on organisms. Thus, intraevolution (microevolution) by natural selection persists within but not beyond species' boundaries. Here we confront the fallacy of the evolutionist assertion that with time all things are possible.

There are no missing links. The scientific data on the platypus and turtle, and the gaps in the fossil record that parallel the gaps we observe in the living world, attest to the illusion of "missing links." Without these putative "links" the Darwinian postulate of molecules-to-human evolution by natural selection (evolutionism) will eventually be discredited. The 24-30 January 2009 issue of *NewScientist*, in its cover story marking the 200th anniversary of Darwin's birth, tendered its farewell to the evolutionary tree of life. The title "Darwin Was Wrong—Cutting Down the Tree of Life" marks the beginning of the end.

Life forms are either extinct, like the dinosaurs, or extant. Dinosaurs are frequently cited as intriguing evidence for the evolutionist worldview. According to such proponents, dinosaurs died out sixty-five million years ago when no grasslands supposedly existed. However, recent fossil evidence shows that dinosaurs ate grass and that the remains of dinosaurs contain soft-tissue, flexible blood vessels, cells, and a collagen-like bone matrix. These empirical facts are contradictory to the evolutionist worldview but consistent with the biblical account in Job **40**:15-24 that dinosaur-like creatures lived contemporaneously with human beings. Such discoveries warrant modifying or abandoning the evolutionist worldview, but instead secondary assumptions are made to accommodate the antithetical evidence.

The creationist/evolutionist controversy began, of course, with Darwin. Contrary to the views of many of his disciples, Darwin did not change science. Science, the study of nature, has not changed, but the scientific community has shifted from a predominantly religious orientation to an increasingly secular, agnostic, and atheistic outlook. Scientists are unanimous about the relevant facts but divided over their interpretation and philosophical implications. Evolution is a fact in the sense that human beings can produce different breeds of animals and plants. Darwinian evolution is not a fact in the sense that human beings through farm breeding or in science laboratories are unable to produce breeds that are intermediate, say, between a cat and a dog. If linearity of the DNA program is what affirms the Darwinian theory of evolution, then it has failed in this regard.

Delusion is the deliberate and continued insistence on the impossible in the face of mathematical, empirical, and objective truth. Darwinism has all the symptoms of an intellectual delusion. The rules of scientific integrity demand that every scientific theory must be compared with alternative theories. Darwinists, however, reject this expectation for fear that the evolutionist worldview will be defeated by the creationist worldview. In fairness to children of any generation, both worldviews must be taught in every educational institution and in all places of worship. Empirical evidence says "No" to the doctrine of life's spontaneous generation. Darwinists say "Yes" and tell folk stories of miraculous primordial conditions that produced living matter at some point in Earth's history. Such imaginary events in the past for which no evidence exists cannot constitute a legitimate part of science. The fossil record says "No" to evolutionism for its failure to demonstrate the numerous transitional stages between molecule and human being, whereas Darwinists say "Yes" and blame the geological record for being imperfect. Information theory, mathematical deductions, and common sense say "No" to the evolutionist worldview because the genetic program is far more sophisticated than any existing computer program. Darwinists, on the other hand, say "Yes" and argue that inert matter can blindly program itself without outside intelligence. If the human race are, indeed, the product of mindless natural processes, then these embarrassing interpretations of the scientific data should be *prima facie* evidence. But, of course, the truth is that human beings are endowed with great and complex minds. Little wonder that a growing number of thinkers, particularly in countries where the creationist/evolutionist controversy still rages, are insisting on the need to restore scientific integrity. Scientific facts are transparent and need no defense, but pseudoscientific speculations and deductions require constant defense in science classrooms and courts of law. It is against the spirit of science to make irrational assumptions, such as abiogenesis, in order to get around evidentiary roadblocks and establish a preferred worldview. If scientists manipulate data to justify a philosophical preference, some people would expect them to do the same in order to achieve a political goal. Public administration's interference in science is one of the reasons why the Union of Concerned Scientists is fighting to restore scientific integrity in policymaking. The glory of science is waning under the mandarin insistence on evolutionism.

Both creationism and evolutionism have profoundly religious implications. The controversy between these two camps will continue indefinitely so long as both sides rely only on pieces of circumstantial evidence. Such fragmentary evidence, however, is inconclusive. God's claim in the Jewish and Christian Scriptures is either true or false and hence conclusive. Natural selection, a mindless process, is useless in this regard. The Judaeo-Christian God intervened to claim credit for creating the world with power, wisdom, and understanding.

Dawkins' *The God Delusion* purposely ignores the crucial historical event of God's claim, for it discredits his argument. His omission is consistent with Dawkins' pattern of entertaining his readers by capitalizing on issues that are seemingly supportive of his thesis while slighting other evidence. For instance, Dawkins devotes many pages of his book to developing an Einsteinian construct of God that favours his thesis, but he sees no need for a Newtonian God because that conception opposes his own beliefs. Dawkins gives the impression that Newton had no other option than to embrace the dominant creationist worldview of his time. Einstein, however, endorsed the idea of a Divine Creator, albeit an impersonal Being. Dawkins also neglects to inform his readers that Einstein never subscribed to the Darwinian theory of evolution by natural selection. Why was Einstein reluctant to endorse the Darwinian theory of evolution?

A solid scientific theory is only tenable for events that are testable and reproducible. For the bacteria-to-bacteria or finch-to-finch evolution, the mechanisms are understood because the events are testable and reproducible. This field of evolution, thus, represents good science and it is not antithetical to a belief in the existence of a Supreme Creator. However, for the bacteria-to-human evolution, which is the essence of the Darwinian Theory of Evolution, the mechanisms are not understood, as the events are not testable and reproducible. Since this autonomous field of evolution is outside the domain of science, acceptance of the evolutionist worldview is based on faith that is clearly antithetical to a belief in a Divine Creator and hence contrary to Einstein's perception of the universe.

On the matter of origins, the appropriate question that every individual must answer is not whether one believes in creation or evolution; it is whether one by faith believes in creationism by divine intervention or evolutionism by chance.

The Vatican deviated from Scripture when, to mark the bicentenary of Charles Darwin's birth, it endorsed the Darwinian postulate that humans descended from apes. The Vatican is essentially repeating the same purblind error it made regarding Galileo. In that case, the Vatican endorsed prevalent opinion and interpreted Scripture to disagree with the scientific fact that the sun as opposed to the earth is at the centre of the galaxy. The Church construed such figurative expressions as the "sun stood still" literally, neglecting other Scriptures such as "moves the earth out of place" and "sits above the circle of the earth." In the case of Darwin, the Vatican interprets God's claim of having uniquely created Adam and Eve as our primordial ancestors to be consistent with the pseudoscientific myth that man and woman evolved from slime by chance. The Vatican thus allegorizes God's literal reference to "six days of creation" as betokening millions of years of evolution. If the Vatican is correct, it must rewrite the Scriptures to explain from which family of apes the first Adam descended. The Vatican cannot on the one hand endorse the Darwinian model of bacteria-to-human evolution and on the other hand continue to honour God's personal claim to have created the world. It therefore owes its adherents an explanation as to why they should continue to honor God's Ten Commandments as holy, righteous, good and true if the Creation Sabbath Commandment, the cornerstone of the Decalogue, is a myth. Why would God claim to have created living things if they evolved with no plan or purpose? It is to discourage conflicting interpretations of this sort that God presented the fact of having created the world as a moral commandment law. Under the evolutionist worldview, there is no rationale for the Atonement. The Vatican is clearly wrong in its assertion that the basic tenet of Darwinism is compatible with Christian doctrine. Either God as the totality of truth created the world with power, wisdom, and understanding, as claimed, or the world came about by chance with natural selection as the sole designing instrumentality.

For mutually exclusive propositions, only one can be true. Unless God's historical claim to have created the world is proven false, then God is no delusion, and therefore any contrary worldview is quite simply the delusion. This book concludes that evolutionism is a delusion in the minds of people who regard themselves as products of natural selection. Such persons choose to ignore the permanent benchmarks of God's ownership of the world: God's six days of creation and one day of rest define our weekly cycle, for which there is no astronomical reason. Moreover, God's intervention through Christ to redeem the fallen human race traditionally calibrates our years of earthly existence into B.C. (Before Christ) and A.D. (Anno Domini). These historical facts and others explored in this book attest to God as the Creator and author of all naturalistic processes.

Moral values as well as belief in creationism or evolutionism are choices we make, but God's existence is not a choice. After 150 years of Darwinism, any continued attempt to replace God with natural selection by using extrapolations to bridge the gap between science (intraevolution/facts) and pseudoscience (extraevolution/myths) will only confirm further the depth of the Darwinian delusion.

ACKNOWLEDGEMENTS

I thank God for the special energy and extra strength that I needed to work through the late hours of the night to compile this book.

My precious wife, Margaret, and daughters, Mary-lyn, Mercy, and Michelle deserve my gracious thanks for their patience, encouragement, and contributions to this book. One night my daughter Mary-lyn came to wake me up at 2 a.m. to go on the computer and continue the work for the Lord. It was a very productive call to duty! Margaret made editorial suggestions and filled in for most of my family responsibilities so that I could concentrate on this book.

I have benefited greatly from the comments, corrections, and suggestions provided by the editor the Rev. Dr. Ayyoubawaga B. Garfour. Upon reflection on his contributions, I can say with confidence that God prepared him for this task.

I must also acknowledge the editorial assistance of edit911. com., Professor Vittorio Frigerio, Carolyn Baxter, Perryne Constance, and Alisha Halkampi. A big thank you to Rob Fredericks and Mary Felice. The discussions and encouragement by Paulette James, M. E. Maingot, Dr. F. Ahiya and Daniel Ahunanya are praiseworthy. Special thanks to the entire Amah, Appah, Bassey, Ebifegha and Fiebai families. My sincere gratitude to John Kucheti, Esther Oduru, Vivian Adeosun, Lanre Adeosun, and Tammy Adoki.

This book owes its existence to the numerous references cited. I extend my special thanks to these authors. My profound thanks to Author House for making self-publishing available so that authors like me can be heard anytime and anywhere.

APPENDIX A

TEACHING OF ORIGINS

The public is given the impression that the creationism-evolutionism controversy is between scientists and religious fundamentalists. The truth is that the controversy is between two categories of scientists who disagree on philosophical grounds. Twenty-seven British scientists and educators signed the following letter addressed to the Secretary of State for Education, Estelle Morris, regarding the teaching of origins, in 2002. Clare Dean under the title "Let's Teach Science Pupils How To Think" reported this letter in the Times Educational Supplement of April 26, 2002.

These specialists in physics, chemistry, biology and geology insist that creationism and evolutionism should be considered side by side in science classrooms since both dogmas cannot be demonstrated empirically. The details are as follows:

The Rt Hon Estelle Morris, MP
Secretary of State for Education and Skills

Dear Secretary of State

Teaching of Origins in Schools

The undersigned academics, scientists and educationists are deeply concerned that the reasonable position taken by the QCA in National Curriculum science and by Ofsted concerning the teaching of origins at secondary level has been challenged. (We write as a group of individuals and consequently the views expressed do not necessarily represent the view of those organisations with which we are associated).

The National Curriculum requires that Darwinian evolution is put across as the dominant scientific theory but also requires that pupils are taught "how scientific controversies can result from different ways of interpreting empirical data". Science should be taught with the critical appraisal of alternative theories. Such debate concerning opposing theories provides rigour in scientific method and contributes to the development of critical thinking by pupils.

We find it most inappropriate that some well-meaning scientists have given the impression that there can only be one scientific view concerning origins. By doing so they are going way beyond the limits of empirical science which has to recognise, at the very least, severe limitations concerning origins. No one has proved experimentally the idea that large variations can emerge from simpler life forms in an unbroken ascendancy to man. A large body of scientific evidence in biology, geology and chemistry, as well as the fundamentals of information theory, strongly suggest that evolution is not the best scientific model to fit the data that we observe.

We ask therefore that, where schools so choose, you ensure an open and honest approach to this subject under the National Curriculum, at the same time ensuring that the necessary criteria are maintained to deliver a rigorous education.

Yours sincerely

Andy McIntosh DSc, FIMA, CMath, FInstE, CEng
Professor of Thermodynamics and Combustion Theory, University of Leeds

Edgar Andrews BSc, PhD, DSc, FInstP, FIM, CEng, CPhys.
Emeritus Professor of Materials Science, University of London.

David Back BSc, PhD
Professor of Pharmacology & Therapeutics, University of Liverpool

Stuart Burgess BSc, PhD, CEng, MIMechE
Reader in Engineering Design, University of Bristol

Sylvia Baker BSc, MSc
Head, Trinity Christian School, Stalybridge

Nancy Darrall BSc, MSc, PhD, MIBiol
formerly Research Officer, Central Electricity Generating Board

Graham Everest BSc, PhD
Professor of Mathematics, University of East Anglia

Ian Fuller BSc, PhD
Lecturer in Physical Geography, Northumbria University

Nick Fuller BSc, PhD
Post-doctoral research (Molecular Biology), University of Warwick

Colin Garner BTech, BEng, PhD, CEng, MIMechE, MSAE
Reader (Applied Thermodynamics), University of Loughborough

Paul Garner BSc, MIInfSc, FGS
Senior Information Scientist, Cambridge Science Park

D B Gower BSc, PhD, DSc, CChem, FRSC, CBiol, FIBiol
Emeritus Professor of Steroid Biochemistry, University of London

Terry Hamblin MB, ChB, DM, FRCP, FRCPath
Professor of Immunohaematology, University of Southampton

Arthur Jones BSc, MEd, PhD, CBiol, MIBiol
Science Education Consultant

Nigel Jones MS, FRCS
Consultant Vascular Surgeon, Freeman Hospital, Newcastle
upon Tyne

Geoffrey Lewis MA, FSA, FMA, HonFMA
formerly Director of Museum Studies, University of
Leicester,
past President, International Council of Museums

Derek Linkens BSc(Eng), MSc, PhD, DSc(Eng), ACGI,
CEng, FIEE, FInstMC
Research Professor in Systems Engineering, University
of Sheffield, formerly Dean of Engineering, University
of Sheffield, past President, Institute of Measurement
and Control

Jeff Lowe MSc, MCGI, DMS
formerly Principal Lecturer, Manchester Metropolitan
University

John Peet BSc, MSc, PhD, CChem, FRSC
formerly Science Coordinator, Guildford College of Further
and Higher Education

David Rosevear PhD, CChem, FRSC
formerly Senior Lecturer, University of Portsmouth

Nigel Robinson BSc, PhD
Post-doctoral research, University of Leicester (Currently
teaching Chemistry)

Stephen Taylor BSc, MEng, PhD, ACGI, MIEE
Reader in Electrical Engineering and Electronics, University
of Liverpool

David Tyler BSc, MSc, PhD, CertEd, MInstP, CPhys, ACFI
Senior Lecturer, Manchester Metropolitan University

David Walton BSc, PhD
Visiting Lecturer, Dept of Computer Science, University of Durham
Information Systems Consultant

David Watts PhD, FRSC, FInstP, FADM
Professor of Dental Biomaterials Science, University of Manchester Dental School

Tim Wells BSc, PhD
Lecturer in Neuroscience, University of Cardiff

Bill Worraker BSc, PhD
Senior (Software) Development Engineer, Hyprotech UK

The above signatories are all accomplished scientists; they are not religious fundamentalists

Reference
http://www.biblicalcreation.co.uk/educational_issues/bcs116.html

APPENDIX B

THE GOD OF CREATION

&

THE TEN COMMANDMENTS

The God of Creation

For the interest of people who are not familiar with the history of ancient Israelites, it is worthwhile to present a summary of the historical events reported in Exodus Chapters 19, 20, 24 and 31 and Deuteronomy 5.

God claims ownership of the universe [Exodus 19:5] and endorses the people of Israel as a special possession subject to their acceptance to remain holy by sticking to God's covenant, which is generally referred to as "The Ten Commandments". One of God's attributes is holiness. God's interest in the affairs of the world, as a stringent requirement, is to be communicated through a holy nation and a cabinet of priests. Israel, among all nations of the world, is God's choice. God briefs Moses on this matter three months after they were delivered from slavery in Egypt. Moses presents God's plan to the elders in the Israeli community: it is accepted and a formal meeting of the entire nation with God is scheduled three days in advance [Exodus 19:7-13]. In the agenda of this transcendental event, God is to address the people from the heart of fire and dark cloud at the summit of Mount Sinai; a long blast from a ram's horn signals when the audience should approach but not touch Mount Sinai.

In the morning of the third day, ancient Mount Sinai blazes and quakes at God's unique presence to address the nation. Thunderstorm and trumpet blast commands attention. Moses leads the crowd to the foot of the mountain. God presents the terms of the Covenant, "The Ten Commandments." The first three exclusively addresses how they should relate with God as the Sovereign Head of the Universe [Exodus 20:1-7]; the last six describes how they should relate with each other as neighbours [Exodus 20:12-17]. Sandwiched between these two sets of commandment, is the Creation Sabbath Commandment [Exodus 20: 8-11], which also carries the dual role of a *sign*, and a unique, and autonomous *covenant* on creation [Exodus 31:12-18]. It establishes how people should affiliate with God as the Creator and treat others including domestic animals as important to God. As a holy nation and a cabinet of priests people affiliated with God should work six days to conform with the six days work schedule God adopted in establishing the heavens and the earth and their contents; the Sabbath day (seventh day), as a seal on creation, is to be honoured and celebrated by abstaining from work [Exodus 31:15-17]. This Creation Sabbath law embodying cosmological information and personal claims is the central focus of this book.

The meeting is orderly and fruitful. However, God's voice generates fear, the meeting concludes with the elders imploring Moses to excuse the mass from any of his meetings with God. Moses stays on the mountain forty days and forty nights (about six weeks) after this human-divine meeting [Exodus 24:12-18]. Moses receives God's Commandment in printed form on two tablets of stone. The historical account in the Scriptures reads, "Then Moses turned and went down the mountain. He held in his hands the two stone tablets inscribed with the terms of the covenant. They were inscribed on both sides, front and back. These stone tablets were God's work; the words on them were written by God himself [Exodus 32:15-16, New Living Translation].

Moses' six weeks absence creates concern and one of the terms of the covenant between the people of Israel and God is broken as they worship a golden calf. Moses in annoyance smashes the tablets on the foot of the mountain [Exodus 32:19-20]. Moses on God's invitation spends again another forty days and forty nights on Mount Sinai after which he gets a second set of tablets containing the words of the covenant—the Ten Commandments. Moses returns to the people this time with both the tablets and a radiant face [Exodus 34: 29-35], affirming the supernatural involvement of his mission. Moses appearance generates fear among the people, he calms them and presents them with the terms of the covenant. With a face too bright for the natural human eye, Moses wears a veil over his face when interacting with people.

The Ten Commandments

The God Creation Sabbath (GCS) Commandment was issued sandwiched between two sets of moral commandments. One set (vertical component) exclusively describes how human beings are to relate to God as their creator and the other set (horizontal component) describes exclusively how human beings are to relate to each other as neighbours. The GCS Commandment is unique from the others because it is the only one that integrates both components. The Ten Commandments are summarised as follows:[1]

1   No disloyalty to God—sacredness of allegiance to God.
2   No misrepresentation of God—sacredness of worship exclusive to God.
3   No disrespect for God—sacredness in the use of God's name.
4   Honour and emulate God as Creator—sacredness of God's Creatorship/Ownership/Sovereignty.
    Respect for neighbours, including domestic animals—sacredness of relationship.
    Honour the Sabbath day—sacredness of time.
5   Honour your father and mother—sacredness of parenthood.

6   No murder—sacredness of life.

7   No adultery—sacredness of marriage.

8   No stealing—sacredness of property.

9   No false testimony—sacredness of justice.

10   No coveting—sacredness of desire.

The entire set of Ten Commandments (or Ten Words) is what is referred to as the *Decalogue*. The full version of the words that God spoke and inscribed on tablets of stone, is available in Exodus Chapter 20.

Moses, in reminding the wilderness generation of Israelites about their unprecedented face to face meeting with God, in the Deuteronomy account, said:

These are the commandments the Lord proclaimed in a loud voice to your whole assembly there on the mountain from out of the fire, the cloud and the deep darkness; and He added nothing more. Then He wrote them on two stone tablets and gave them to me. When you heard the voice out of the darkness while the mountain was ablaze with fire, all the leading men of your tribes and your elders came to me. And you said, "The Lord our God has shown us His glory and His Majesty, and we have heard His voice from the fire. Today we have seen that a man can live even if God speaks with him. But now, why should we die? This great fire will consume us, and we will die if we hear the voice of the Lord our God any longer. For what mortal man has ever heard the voice of the living God speaking out of fire, as we have and survived. Go near and listen to all that the Lord our God says. Then tell us whatever the Lord our God tells you. We will listen and obey." The Lord heard you when you spoke to me and the Lord said to me. "I have heard what this people said to you. Everything they said was good. Oh, that their hearts would be inclined to fear Me and keep My commandments always so that it might go well with them and their children forever! Go, tell them to return to their tents. But you stay here with Me so that I may give you all the commands, decrees and laws you are to teach them to follow in the land I am giving them to possess." [Deuteronomy 5:19-24, NIV].

The relevance of providing both oral and written testimonies in claiming credit for creating the universe is noteworthy. While the original inscription by the finger of God can either be misplaced or destroyed by human beings or retrieved by God, the oral testimony cannot be misplaced or destroyed by humans or revoked by God. In fact, out of annoyance with the congregation of Israelites over their worship of the golden calf, Moses destroyed the first set of tablets God wrote upon and had to solicit for a second set. The interested reader should read Deuteronomy 9:7-29, 10:1-5.

INDEX